A SMALL TREE IN A TEXAS HURRICANE

A Memoir

Robert Henry Benson

ISBN: 979-8-9952798-0-8

Published by Pauraque Ridge Press
Normanna, Texas

Printed in the United States of America

First Edition

For Karen, who has been the still point at the center of
every storm I made.

Success is not measured by what a man accomplishes, but by the opposition he encountered and the courage with which he has maintained the struggle against overwhelming odds.

Charles Lindbergh

Author's Note

The events in this book are true to my memory of them. I have lived with these stories for decades, turning them over the way a man turns a river stone in his hand, worn smooth by time, but real. Memory is not a courtroom transcript, and I make no claim that every conversation appears here word for word. What I do claim is that the feelings were true, the moments were real, and the people who shaped me were as I have described them.

Some names and identifying details have been changed to protect the privacy of individuals who never asked to appear in anyone's book. A few characters are composites of people I knew during the same period of my life. I have done my best to be fair to everyone on these pages, including myself, which was sometimes the harder task. This is one man's account of one man's life, told as honestly as I know how.

CHAPTER 01

Otto's Tamales

"You got to taste some bad ones to know the good ones."

Otto Galvan said this to me on a lunch break in August of 1969, sitting on an upturned bucket in the heat, unwrapping a tamale his wife had packed. He said it the way a man states a fact he has known so long it no longer surprises him. He wasn't talking about tamales, not really. He was talking about everything.

I was twenty-seven years old. I didn't understand him yet.

We had spent the day lifting 37-pound concrete blocks onto a two-story wall at Southwest Texas State University in San Marcos, Texas. By quitting time, my jeans and steel-toed boots were caked with mortar, mud, we called it, and even my hard hat carried its share of splatter. The older men on the crew called the blocks "birth control blocks." A day of lifting them dissolved every thought of procreation. All you wanted at the end was a hot shower and a bed.

Otto was twice my age, dark-skinned, born in New Braunfels to a Texas-German mother and a Hispanic father. He spoke German, Spanish, and English, the first trilingual person I'd ever known, and somehow managed to carry tamales in his lunch box no matter the season. His knee had been destroyed by decades of lifting. By the end of a day, it would hurt him badly enough that he needed help getting down from the scaffolding. I almost always climbed back up and grabbed his tool bag, hauling it down so he could use both hands on the hot metal frames. He always thanked me with a sweaty handshake.

I enjoyed his company. During our thirty-minute noon break he would talk, and I would listen. Otto was a salt-of-the-earth philosopher, the kind of man whose sentences stayed with you after the conversation ended. More than once, holding a tamale from his wife's kitchen, he would say, "You got to know, young man, what makes these so good is having tasted some terrible bad ones in your life. Mi esposa don't make the bad ones. You got to taste some bad ones to know the good ones."

His face told me what my future might look like if I stayed on the scaffold. Worn down. Body broken. Spirit dulled by heat and repetition.

I held my own tools the way a man holds things he owns: a canvas bag worn soft by years of use, carrying a trowel, a fold-up ruler, and wooden line blocks that still smelled of lime and sand. In my right hand, a four-foot bubble level with brass edging, expensive, and mine, a reliable partner through each day's grind.

When we climbed down that Friday, I stepped off the limestone campus and onto North Street, where the tidy university gave way to dilapidated frame houses, overgrown yards, and peeling paint. My car was parked three blocks away.

At the crosswalk on North Street, a young teacher was shepherding a flock of uniform-clad second graders across the intersection. She looked about the same age as most of the students at SWT, but her dress was professional, conservative. She smiled at me while I waited for them to cross.

The last kid to cross was a boy with blondish hair and both front teeth missing.

I stopped walking.

Back in my second grade, I had been that boy. The one who lagged. I'd had the same gap where my front teeth should have been. And just like that, standing on a sidewalk in the August heat with mortar on my boots, my mind flooded with a memory I had spent years trying not to revisit.

Her name was Sarah. She sat at the desk in front of me.

I couldn't write well in cursive. My letters came out wrong, looping in the wrong directions, refusing to connect the way Mrs. Lawson drew them on the board.

Sarah knew this, and she used it. Every time Mrs. Lawson turned to the blackboard, Sarah would swivel in her seat and deliver her verdict in a mean whisper:

"Curly letters, twisty line, Bobby's always last in line."

I bore it for weeks. I don't know now what I was waiting for. Some limits I hadn't found yet, some final repetition that would finally be too many. It came on a day I can no longer place, a normal school morning, nothing remarkable about it, when she started up her chant again and something in me simply stopped waiting.

I lifted the small jar of blue-purple ink from the well on my desk, stood up, and poured it directly on top of her head. The stain ran through her hair, into her eyes, onto her neck, and down the front and back of her clothes.

Mrs. Lawson grabbed my arm before I could appreciate what I'd done. I remember the glass jar falling from my hand, I can still see it tumbling in slow motion, the tiny membrane-like cup near the top catching the last drop of ink. Then the principal's office. A paddle with holes drilled in it. Three licks that burned. Sitting in the hallway while my mother was called.

On the ride home, she didn't speak. I looked straight ahead and felt sadness.

That might have been the end of it. But the last day of school arrived, and every other child received a report card. I did not. Mrs. Lawson asked me to wait until everyone had gone.

"Bobby, sit down and let me explain something to you."

She told me I had not met her standards. I would be repeating third grade in the fall.

My mother was called again. She didn't speak on the way home. I looked straight ahead again, this time with something heavier than sadness, a feeling I didn't have a word for yet, sitting in my chest like one of those concrete blocks.

Later that night I lay in bed unable to sleep. My mother had a neighbor over. I could hear their voices from the kitchen, the soft sounds of two women drinking coffee and smoking Chesterfields at the table.

I heard my mother begin to cry.

And then her words came through the dark, not meant for me, not meant to be heard:

"Doris, I wish Bobby had been born smart like his brother. Johnny is so good and never gets in trouble. All his teachers say nice things about him. But Bobby, he's not the same. I don't know where he'll end up, maybe at the same hospital as his father. If he could sit still, maybe he could learn something. I don't know what's going to happen to him, and it hurts my soul."

I withdrew to a secret place deep inside myself, where the sound couldn't follow.

Her words didn't make me angry. They made me believe her.

I was eight years old.

Standing on that North Street sidewalk in 1969, watching the gap-toothed boy disappear around the corner, I felt the distance between that kitchen and this scaffolding like something physical. Sixteen years. Same boy, different boots.

Otto's tamales were packed away. The crew had scattered to their cars. The Texas sun was dropping toward the hill country, painting the limestone buildings in the kind of light that makes everything look temporarily valuable.

I picked up my tool bag and started walking.

I didn't know it yet, not fully, but something had shifted in me that week. The scaffold. The men with their ruined bodies. Otto and his bad tamales and his good ones.

The distance between where I'd come from and where I might go was, I was beginning to understand, exactly the point.

CHAPTER 02

Deco's Porch

I'd parked my 1965 Volkswagen Beetle three blocks away, in front of one of those weathered houses. The paint had faded and the body showed its years, but it was mine and paid for, even if its latest repair had cost more than the car was worth. It sat in front of a house that had stopped trying. We had that much in common.

The young teacher I'd just watched crossing the street had already disappeared around the corner. My marriage was on the rocks. I should tell you how she came into my life.

I was twenty-one and working in Gastonia, North Carolina, helping build a masonry warehouse. At a local bar, I met Rachel Smith, three years older than me. She worked as a night-club dancer. After a whirlwind courtship, we ran to South Carolina, where a couple could marry without delay, and tied the knot.

Rachel's life experience was broader than mine. She had three kids from a previous marriage, but for reasons I never knew, her children were in foster homes. I was infatuated with this older woman. I was shy and uncomfortable approaching women I didn't know. Rachel singled me out and showered me with attention. I was far from Texas, lonely and unaccustomed to female affection. If Rachel had a plan for me, it worked.

One night, after we'd been married only a few months, Rachel sat on the edge of the bed, sipping her Cuba Libre. She looked straight at me and said, "You know why I married you, don't you?" Her voice carried no warmth.

"It wasn't for love. I needed a husband. The county court is not going to give me my kids back unless I have a husband. You were young, stupid, and you wanted me."

She drained the glass, set it hard on the nightstand, and added, "This was about my kids, not about us."

I sat frozen, staring at the floor, the ice cubes still clinking in her glass.

Our marriage might have worked if I'd been more mature. We wanted different things, and my restlessness was her final straw. She filed for divorce in the spring. I can say one good thing about that marriage: we made a daughter who grew into a fine woman, and I am proud of her.

The divorce would add to my financial troubles. Now I'd have child support to pay on top of all the other expenses of living. I needed an extra job. Unless I found a nighttime or weekend job, I'd be finished, maybe in jail for nonsupport. I sifted through my limited abilities and considered what might get me an extra job. Options were scarce. Brick masonry was my only real skill.

My steps quickened. Two days off. I lengthened my stride, thinking about lying down on my worn carpet and not moving until Monday.

Before I reached my VW, I noticed a guy sitting on the steps of the shabby house. He was skinny, with a reddish beard and matching hair. He wore glasses with heavy black rims and was smoking a joint. He stared at me as I approached.

"Hey, you." He was looking straight at me. "You took up my parking space all day. I hope you can get that piece of German shit started."

I was in no mood to argue. I equaled his stare but didn't say anything for a long moment, hoping my silence would put an end to the exchange. He seemed to take that silence as an opening to go on.

"Hey, man. I'm just pulling your leg. Anybody can park where they want around here." He shrugged. "I just usually get that spot. What do you do around here to make you look like that?"

"Look like what?" I asked.

"I mean, you got shit all over you," he said. "Have you been working with cement or something?"

I tossed my tools in the trunk and didn't answer right away. I still had to drive to Seguin. Seguin is where I rented a small apartment.

I took a deep breath and turned around with the best smile I could muster. "I'm a bricklayer with the crew building a utility barn at the University. Laying blocks. Heavy as Hell. You don't feel much like talking if you do that all day. They won't let us park near the job. Cars have to be parked off campus. Sorry about grabbing your spot."

He took another drag on his joint and squinted. "It's more than 25 miles to Seguin. How many days are you going to be working up here?"

I watched him slowly exhale the smoke, which drifted around his unkempt hair. "I think the job will be finished in about two weeks. It depends on whether any old guys quit because they have to lift those blocks on their own. Most outfits say we should pair up when we need to lift blocks above our waists, but the boss on this job won't allow it. Some of these guys can't do it like that and just quit. Then the job takes longer."

"Tell you what. I go to school here. I'm a grad student in the Business Department. I'm sort of the head of this old house because I've been renting here the longest. It's not pretty, but there are four bedrooms, two bathrooms, and a kitchen. One of the bedrooms is vacant. I'd rent it to you cheap, if you don't want to drive from Seguin every day. How does ten dollars a week sound?"

I hated that drive from Seguin. The traffic coming into San Marcos in the mornings was terrible, and the same in the afternoons. Besides that, it was tough to find a parking spot. I didn't like the look of this guy on the steps, but ten dollars a week was hard to argue with.

"I might take you up on that. My brother goes to school here, so I'd know somebody. You going to be here Monday morning?"

"Dude, except for going to one of my classes when I feel like it, I'm always here. But I don't get up too early. Morning sounds way too early for me. I usually sleep in. It'd be better if you came on Sunday afternoon." He laughed at his own schedule.

He offered me the last drag on his lip-wetted reefer. I refused. I didn't like getting high, and I had a standing rule about things that had touched someone else's mouth. Cigarettes, reefer, didn't matter. If I saw it happen, I couldn't take it.

The guy on the steps had unruly red hair not just on his head, but on his skinny arms. Seeing someone with red hair always split me in two directions at once. Redheaded girls were magnetic, almost intoxicating. Shirley McGlothlin, with her freckles and copper waves and mischievous smile, had been my first girlfriend in junior high. But not the redheaded man from my childhood — the one from the asylum laundry. That memory repulsed me entirely.

"Listen, man, I've got to get home now, but I'll think about your offer, and if I decide to do it, I'll come back Sunday evening with some clothes and things. Are the other guys that live here like you?"

"There are two of them, both upstairs. Jeff is a drama major, and Paul is in the history department. I'm Declan. Declan Reid. They call me Deco for short. My room is next to the kitchen downstairs. The empty bedroom is next to the downstairs bath. We'd be sharing that bathroom." Deco moved his roach to his left hand and extended his right hand to shake mine. What a different world that guy Deco lived in. I wasn't sure his future in twenty years would be much better than mine if he smoked dope all the time. I couldn't see why he was so open about it, given that it was illegal. You couldn't even buy a beer in Hays County in the 1960s.

I climbed into the Beetle, mortar dust still itching in my shirt, and glanced back at Deco on the porch. He was grinning like he knew something I didn't. Maybe I didn't belong on that scaffold anymore.

CHAPTER 03

Hey, Kiddo

I didn't know it then, but the path I was on led straight through some shadows I never escaped. One thing was sure: paying rent in two places was foolish. I tried to convince myself it might make sense. Maybe I'd save some gas money, since driving back and forth between Seguin and San Marcos cost something even in those days. But gas in August of 1969 was only about thirty-five cents a gallon, so that excuse didn't hold much water.

I told myself I could spend more time with my brother John, but he was busy with school, his head in solving physics problems, while mine was baking in the sun on a block wall. After laying hundreds of blocks in the swelter of a Texas summer, I didn't feel like socializing much anyway. My life after work was to shower, eat, and collapse, not put on a clean shirt and party.

A wife divorcing me, a little girl who deserved a father who didn't vanish, and now a young woman I'd met who wanted to move in with me, was too much stress. How was I supposed to balance all that? One apartment was already too much to carry; two rents would be plain reckless.

I couldn't imagine myself fitting in with the groups of young people laughing under the live oaks on campus. I had a heavy redneck accent and couldn't remember ever making a grade above a C before dropping out of school.

I was the restless boy who couldn't sit still or finish what he started. But somehow, I still felt the pull to try things. College was not for me. It made no sense, yet there I was, drawn like a fly to an emptied beer can, ready to dabble for at least a few weeks to see if I could fit into this world. I could keep most of my many quirks and oddities hidden, but the one I couldn't keep secret was my aversion to food tasted, touched, or handled by others. Some fears you can stuff down; others push their way up, and mine always seemed to come back to the table. Maybe I was born twitchy, but I preferred to blame my fate and the crooked path of my childhood.

One of the earliest and darkest of my memories came when I was four or five. At that age, children ought to be chasing fireflies in the yard, not worrying about whether their parents could feed them. But life rarely behaved the way it ought to. My father was in the "hospital." At least that's how my mother referred to it. My father was a prisoner in the Texas Prison System.

Mother was working two, sometimes three jobs, never making enough to keep my brother and me with her. She made the hard choice to send us to our grandparents. I understood it even then: she couldn't hold everything together. That didn't stop me from feeling abandoned. For over a year, we lived under my grandparents' roof, and though they provided structure, the price was steep. Their house stood in the shadow of a place that would haunt me for the rest of my life: the San Antonio State Hospital.

They used to call it the Southwestern Insane Asylum, and even as a boy, I sensed the cruelty of the place. Patients were locked up like convicts, many of them without trial or appeal, and some were never told why they were there at all. The courts sent them away, and the doors closed behind them forever. On maps, it looked like a hospital, but in spirit, it was closer to a prison camp. Six hundred and forty acres of land on the south side of San Antonio, intended to be self-sufficient, required every patient able to work to do so. There was no such thing as rest or recovery in that system, just endless labor and the slim hope you wouldn't be beaten for failing at it. Old timers said the ghosts of those who died there still roamed the wards, rattling in the iron beds and echoing in the stone hallways. For me, those ghosts arrived in the night air, carried on the south wind to my grandparents' porch.

My grandparents themselves were hard people, though not unkind. They had come of age during the Depression, which meant they measured everything, food, money, even affection, like it might run out. I admired them, but I hated sleeping at their house. I would lie awake for hours, listening. When the breeze shifted, it brought the sound of screams and wails from the asylum. They were not the cries of one person but a chorus, fractured and wild, seeping into my imagination until I could hear words in them. Maybe it was only the wind. Perhaps I invented it. Either way, it carved itself into me. Fear was my constant bedfellow there, and it made sure I never forgot the place.

The patients endured more than screaming fits. They were slapped, strapped, and shocked with currents meant to drive out demons. Men and women alike were put to work: scrubbing floors, hauling water, hoeing weeds, picking cotton in fields where the sun boiled the earth. That was called "therapy." My grandfather worked among them every day, though not as a patient. He was the hospital's blacksmith, shoeing mules, sharpening plows, and repairing tools. His shop was one of the few places that made sense to me. I loved to stand near him, watching as belts and pulleys spun from a single roaring motor. He'd use a long pole to shift the belts, sending power to a hammer that beats hot steel into shape. There was order in that rhythm, a clear beginning and end. I sometimes wondered how he could walk into that hellish place, out of his neat little world, and face the chaos of that asylum. Maybe he tuned it out. Perhaps he pretended, like I did, that he wasn't really there at all.

My grandmother, Mimmy, wasn't so lucky. She worked in the hospital laundry, one of the most dangerous posts. Day after day, she fed wet sheets into a machine that looked like it had been designed in hell itself: the mangle. Picture two enormous, steaming rollers, jaws wide open, waiting to swallow anything that came close. She and another woman stood side by side, snatching sheets from the wash and guiding them into that machine. The heat was suffocating, August in Texas multiplied by ten. Her hands moved quickly, too quickly one day, when her finger caught between the rollers. The machine grabbed her without mercy, pulling her arm toward its gut. It would have crushed her whole body if her partner hadn't slammed the emergency stop. The scars on her arm never faded. She wore them like some women wear jewelry, an emblem of her endurance. She didn't boast, but she didn't hide it either. It was proof that she had survived the beast.

The very word "mangled" seemed to come from those machines. My mother once told me about the workers who weren't so lucky, men and women pulled in past saving, their bodies flattened like the sheets they pressed. She described it so vividly that I imagined their funerals as if they were sailors given burials at sea, wrapped in white cotton and released to the asylum's winds. My mind drew comic-book panels of them, cartoon outlines folded and creased, carried into the haunted buildings where they joined the chorus of ghosts.

My imagination was too fertile for my own good, but the asylum didn't care. It gave me material to last a lifetime.

Mimmy eventually returned to the same laundry, scars and all. Sometimes she took me with her, telling me to sit on the cold floor with my toys. I remember one day, hotter and more oppressive than most. The room smelled of bleach and wet concrete. The machines pounded and hissed, drowning out any other sound. I sat in a corner with two little toy cars, rolling them back and forth, trying to ignore the weight of the place. Fear always clung to me in the laundry, the fear that anything might happen. That day, bad things did.

I saw him before my grandmother noticed him. A man in a white uniform shuffled toward me, knees bent, arms dangling, moving like some broken-down ape. His frame was so thin I could see the bones in his wrists. His hair stuck out in a wild red mess, and snot ran down his nose. In his hand, he clutched a sandwich, soggy and misshapen, peanut butter and grape jelly leaking out in streaks of purple and brown. He tore at it with his teeth as he came closer, never looking away from me. I froze. My voice caught in my throat. Then I screamed.

The machines roared so loudly that my grandmother didn't hear. He kept coming, mumbling, chewing, slobber dripping from his chin. His eyes were pink and feverish, the eyes of a man who had seen too many ghosts. The sandwich swung toward me, closer, until I could smell its sour, sticky odor. I tried to run, but my legs refused. My body locked up, trapped in the nightmare I'd always feared would step off the walls of that place and into my life.

"Hey kid, kiddo, kid. You hungry? You hungry?"

The words came out between chews, the sandwich thrust toward my face. I clenched my lips tight, turned my head away. He smeared the jelly across my ear, hot and wet. My stomach lurched. I screamed again, hoping Mimmy would appear. At last, she and her co-worker saw what was happening. They moved fast enough to matter, but not fast enough for me. For a moment, I thought they were as used to this as folding sheets. The other woman grabbed his arm, yanking him back.

"You get off that boy NOW, Harold, or I'll turn you in! Get off him right now!"

He backed off, muttering apologies, still clutching the ruined sandwich. "Sorry, sorry. Maybe kiddo was hungry. Don't tell on me, don't tell. I go back now, no kid, kiddo. Okay?" He shuffled away, still bent, still chewing, disappearing through the doorway like some nightmare retreating into its cave.

Mimmy's voice followed him, sharp and steady: "Harold! You get out of here! I'm telling them you were down here where you don't belong!"

And then it was over. I sat frozen on the floor, too shaken even to cry. The smell of that sandwich lingered in my nose, in my hair, in my skin. I could feel it on my ear, sticky and foul. From that day forward, I could never look at a peanut butter and jelly sandwich without seeing his face, without smelling the sour breath and the heat of the mangle. Food touched by others had always troubled me, but after Harold, it became impossible. My stomach turned at the thought. I learned to say no, to turn away, to keep my distance from anything handled by anyone else.

The asylum left me with scars of my own, though mine were invisible. They shaped the way I saw the world, the way I trusted, or didn't trust, people, the way I fit, or didn't fit, among others. Even years later, whenever I tried to set foot on a college campus or sit at a dinner table, those shadows trailed me. The echoes of screams in the night, the sight of my grandmother's mangled arm, and the smell of Harold's sandwich followed me like ghosts, reminding me I was never quite like the others.

And yet, even as those ghosts kept pulling me back, something inside me pushed forward. Maybe it was defiance. Perhaps it was desperation. I didn't know where I belonged, but I knew I had to keep moving, had to keep testing the world to see if there was a place that would hold me. The wall I laid blocks on by day, the campus I tiptoed into by night, the women and children whose lives brushed against mine, they were all part of the very search.

I didn't know it then, but the search was beginning. And the next step, the one waiting, would test me even harder than the asylum ever had.

CHAPTER 04

The Highchair

Have you ever tried to reach backward through your mental fog and touch the first thing your mind kept for itself, not a photograph someone labeled for you, but a moment your own memory chose? I must have been two, or maybe a bit older. It is not a cake with candles or a favorite toy. It is a kitchen, a highchair, a saucer of gray mush, and a noise so thin and sharp it still rings behind my ears.

I am locked in a highchair. The buckle crosses my belly and clicks. The tray presses my ribs. The saucer sits at the far edge, a small moon of dull food with slick skin. I clamp my mouth and shake my head. The sound that comes out of me climbs into a steady wail that fills the room.

My mother sits across the table. Her face is slack with the kind of tiredness that does not sleep away. She tries to quiet me. She coaxes, then pleads. Her words slide past. On the other side sits my father. With each notch, my voice rises, his shadow grows. One hand curled into a fist. His eyes darken. Patience leaks out of him like water from a cracked pail.

He leans forward. The room tilts toward him. The sentence he throws is the first complete sentence I ever remember, and it breaks like glass in the air.

"God damn it, Florine, shut that squalling kid up. Shut him up NOW. I can't stand that little bastard's bawling. Either shut him up, or I'll show you how to get that food down his throat and kick your skinny ass while I'm at it."

My mother's eyes begin to shine. Her mouth closes into a thin line. She is as frightened of James Albert Benson as I am. His voice shakes the plates. My own cry goes higher and then, all at once, cuts to nothing. If I drew in one long breath and held it until the edges of the room sparkled, the balance would change. Adults panicked when the noise stopped.

I fill my lungs. I clamp my lips. I stare at the saucer until it blurs. The silence grows heavy, then heavier. My mother recognizes the sign. Her chair scrapes. Her fingers tug at the strap and free me. When she lifts me, my shin bangs the tray, pain ringing up to my hip. She drags me to the sink and twists the cold handle all the way. The faucet coughs and then throws a bright stream. She forces my face into it. The shock breaks me. I gasp. Water stings my eyes and floods my mouth. My legs kick. My arms thrash in the air, but I cannot escape.

"Oh my God. Oh, dear God, he's going to die this time."

In 1946, in a house like ours, there were no gentle tools for raising an unruly child. There were switches cut from a willow that raised red ladders on the skin. There was a leather belt that sang the same low note every time it struck. Time out was what a referee called. Standing in the corner meant putting your nose to the wall and counting your sins.

Why are these childhood memories haunting me now? These memories stand at the head of a hall lined with doors I have no wish to open. Some belong to people who cannot answer for themselves. Some are too raw to tell. It is enough to say the breath-holding was not the worst thing that happened, only the first my mind chose to save whole. I do not know why. My memories rise like fish I did not bait. One moment, I'm pouring milk over a bowl of Wheaties; and next, I'm two years old, with my face under cold water. I needed a place to be safe. A place where I could think. A place where I might work out some plan to steady me. There was only one place like that I knew.

My grandparents retired to a small house they built near Dewville, in the quiet corner of Guadalupe County, where pasture meets sky. The land was forty acres of sand and post oaks, laced with gray arms overhead. Lakey

Road carried my grandmother's maiden name on a metal sign. When I say I loved their place, I mean it was the one square that always felt safe. The air smelled of hay and turned soil. Light slid through the trees in bands, and the days unrolled more slowly there.

Most of my life had been a mess, and I knew some of it was mine to own. Sometimes I thought my brain came from the factory with pieces in the wrong slots. Sometimes I thought the fault was my father's, the way he laughed and then struck, the way he treated my brother, my mother, and me as targets on a range. The truth was likely both my odd wiring and his violence braided together and tugging me off any easy path.

I decided to take the long way to Dewville. It was a scenic route that avoided Seguin altogether. A way that I could stop at the side of the road and see things with my binoculars. I tossed clothes into a paper bag, grabbed my binoculars, and drove. Finding birds gave me a quiet place I could not find anywhere else. Spotting a species I had never seen, looking in my book to name it, then watching it for a minute, felt like a small victory I could keep. Along Highway 80, the San Marcos River shows itself, and trees gather like family at the banks. Those places call me.

My first stop was the south side of Martindale, where the river folds over a low concrete dam and turns into a bright hiss. Water there is lively, aerated, and quick. Fish crowd the pool. Big trees lean over it, casting long sheets of shade. I parked at what is now Allen Bates River Park and raised my binoculars. The morning was clear. Leaves flashed their silver undersides in the breeze. A Killdeer scolded along the gravel edge. A Great Egret stood with its neck curved like a letter. Above me, vultures rode the air. A kingfisher stitched the pool with a sharp call. Across the road, mockingbirds flicked their tails. Bluebirds flashed color between posts. I found a Red-shouldered Hawk. Its stare pinned me the way raptors always do. I have never lost my curiosity about the plain hunger in a hawk's eye.

As a boy, I did not know anyone else who watched birds. The field guide I borrowed in junior high was written for the eastern states and left too many of ours unnamed. Later, I learned the same author had written a book about Texas. I bought it with money from bricklaying and held onto it like a key. With that book, I could name the birds in and around Seguin. It felt as if someone had finally drawn a map of a country I had walked through for years without a name for it.

Back then, I never thought of writing any of this down. The idea of a memoir came later, in stray talks with my brother John. We joked about our childhood with the kind of laughter that keeps you from choking. When something in our lives went right, we would say, "This is a long way from Leesville." Those years near the tiny town of Leesville were the worst. But maybe the worst was somehow good. If all your days are flat, you never feel the rise when good ones come. The distance between the low days and the high days is how to measure joy.

John and I had a younger brother, Mark, and there was one son earlier than us. He died before we were born: four boys, no girls. Knowing my father, you could imagine he ordered it that way, four canvases to carry his marks. People now call this toxic masculinity. In those years, no one named it. They watched boys turn into men and hoped the parts that hurt would not pass along.

Telling a life like mine in straight order would miss the shape of how memory moves. A smell will drop you through a floor to another decade. A sound can open a door that you have boarded shut. The best I can do is weave the past through the present so you can see what cut the cloth I have worn. The highchair is one thread. Dewville is another. There is more when the time comes.

I left the river and drove on. The road ran past low fences and ponds that shone only after rain. Gates were tied with frayed rope. Mailboxes leaned, names scoured by the sun. I drove with the window down, letting the air move through the car. A scissortail crossed a pasture, its tail opening like a pocketknife. A crested caracara stood on a fencepost as if weighing a case. Cattle lifted and lowered their heads, dark boats on the grass.

As the miles unrolled, I thought about the way I had always fought the sight of food handled by other people. It would be clean to say it all began in that highchair, and perhaps it did. The truth is that my mind stacks episodes until they feel like a wall. When I was a schoolboy, that fear sharpened. I did not know it was sharpening. I only knew my stomach turned when another child's mouth came near my plate. You could call it a quirk, but a quirk is a small thing that passes. This did not pass.

I turned onto Lakey Road, and the world quieted another notch. Ditches were thick with Johnson grass. Grasshoppers sprang in front of the car and vanished sideways. My grandparents' house sat back from the road, a low place under a large Catalpa tree. My grandmother, Blanche Lakey, before she was a Palmero, had inherited the forty acres. My grandfather had hands that could square a corner and set a beam. Together, they built a house that felt sound.

On that land, I learned what peace feels like in a body. It is the way your shoulders drop without you even realizing it. It is the way you do not brace before a door opens. It is the way you sleep without running. I chased fireflies and let them go. I carried buckets and learned the weight of water. I listened to my grandparents talk in the easy rhythm of people who save their breath for what matters. They were not rich. They were not schooled past the necessary. They were, as people say, the salt of the earth, and I loved them for it.

I parked and sat for a minute with my hands on the wheel. The air smelled like dust and warm leaves. A cow lowed beyond the trees. If you had asked me to name the one place I felt loved without price, I would have pointed through that windshield. This place kept me from breaking clean through.

I was not innocent within myself. I had a talent for making trouble, for choosing the hard way while the easy way waited with its handouts. I could blame my father for some of that and often did. I could blame my own mind for the rest. The truth is, you live with what you are given and what you make of it, and those two twist together until you cannot tell one from the other. On the best days, standing beside my grandfather and his workbench, I felt the knot loosen. At worst, even Dewville could not work the tangles free.

From the porch, I saw the thick trunk of the largest post oak, the one my mother called the grandfather tree. Beyond it was a pasture that ran to a barbed-wire fence. Grass lay in waves where the wind pressed it and then let it rise. The boards were warm under my feet. I knocked, then let myself in. I put my bag in the small bedroom off the hall. That house ran on simple talk and on work done as needed, nothing wasted.

At the table later, I turned the question of memory in my hands. Why the saucer and not a toy? Why the snap of a faucet and not cinnamon toast? Some people remember a song. I remember cold water and my mother's voice when she thought I might die. My mind kept the first moment it knew fear and made a copy to carry forward.

If I stood where the river crossed the dam, I could believe I was a man who had been spared. If I sat with my grandparents, I could feel I had been chosen. The truth lay between. I had been both spared and chosen, hurt and held, and I was still trying to fit the parts together.

Before I turned in that night, I stepped outside again. The sky above Dewville was a black plate pricked with stars. A barred owl called from Willow Creek, and another answered far off. When I closed my eyes, the sound carried me back to the San Marcos River and the hawk crossing the basin. It took me forward, too, though I did not know it, to a lunchroom that smelled of bleach and tuna and hot metal trays, to a yardstick tapping the rim of a plate, to a voice that left no room for argument.

What I mean is this. The highchair taught me that my body would fight what it could not stand. The years after taught me that the fight would keep finding me. I told myself I was going to Dewville for quiet, and I did find some. The trouble with quiet is that it lets the old sounds rise. By the time I lay down, the room was dark, and the fan was pushing slow air, and I knew sleep would not land easily.

I watched the fan's shadow turn on the ceiling. The house made its nighttime noises, boards settling, a soft tick in the wall, a car far off on the road. I tried to sort the choices waiting in San Marcos, on a path different from the one my father walked. Hope flickered in me, small but bright. Fear stood beside it and would not budge.

Still, the decision gnawed at me. Was I being reckless, selfish, or even foolish? Did I really fit in at college? Sleep would not come. Instead, the past loosened its catch and began to spill. My memories of childhood stood in front of me as a giant tissue paper ripped into fragments by a 12-gauge shotgun. Just awkward pieces dangling randomly. Most of those pieces were not pretty. I was haunted all that night as those fragments floated by in a foggy pool. But to know me, tell them now.

CHAPTER 05

Legs in a Pan

Unable to sleep in my grandparents' spare bedroom, I heard what seemed like frogs croaking outside. But that was unlikely because frogs don't croak in January. At least I don't think so. Whatever it was, it flashed into my brain one of the few good memories from my childhood. That memory was of a night with my father, a memory that sits alone in its strangeness, one of the few evenings when he wore a grin instead of a scowl. He came through the door holding a long pole with three sharp prongs bound together at the end. To me, it looked like a weapon, a miniature trident, and my stomach clenched. His moods were storms I had learned to fear, and anything sharp in his hand could mean pain.

"Hay, Bob-a-re-Bob, let's go to old man Jamison's cattle tank and gig some bullfrogs. We'll chop off legs and fry them up for supper. You ain't had nothing in your mouth yet as good as frog legs battered up in corn meal and fried in melted lard."

The words chilled me. Gigging meant nothing to me but chopping and frying frog legs sounded like cruelty waiting in the dark. I searched his face for the anger I usually found there, but instead his eyes glittered with excitement. His grin stretched wide, almost boyish. I thought of the garden snake he'd split open with a shovel and made me touch, or the firecrackers he'd set among ants to see them scatter. Was this another trick, another test of obedience?

"Daddy, I don't want to. I don't want to cut the legs off frogs and cook them. I don't like frog legs, Daddy."

He only laughed, throwing his head back, the sound booming off the walls. "You're coming with me," he said. "You need to learn how to catch your food if you ever find yourself in a pinch. Go grab your shoes and meet me in the car."

"Are Mommy and Johnny coming too? Don't they need to know how to catch frogs too?"

"No. Your Mommy is going to stay here and heat up the grease so those legs will start twitching when she tosses them in the pan. Johnny is just a baby. Too little to go. You get up front with me. I need my headlamp so we can blind'em before we stab them. They won't even know what hit them. I'm telling you, those frogs are good eating. I promise."

Outside, the night pressed close. Trees arched over the road, their branches clawing in the dark. The moon slipped behind thin clouds, then returned, painting the ground silver for a moment before fading again. The safety of the house dropped behind us, and we drove into the thick air of spring.

At Jamison's gap we stopped, slipped the barbed wire loose, and eased down the dirt path to the tank. Frogs called from every direction, the sound swelling and falling, a living chorus. My father carried the gig in one hand and adjusted the carbide headlamp with the other. Its pale beam cut through the shadows, catching eyes that shone like sparks at the pond's edge. The lamp hissed softly, burning acetylene, a relic from miners' days. To me, it was just another piece of strangeness, a ghostly fire strapped to his forehead.

He whispered that the low bellowing, the "RUM, RUM, RUM" that sounded like cattle in the dark, belonged to the most enormous bullfrogs. Those were the ones worth the trouble, thick legs ready for the skillet. I shivered, hearing them call again, the sound vibrating in my chest.

He froze suddenly and squeezed my shoulder. Twenty feet away sat a frog that looked monstrous, its throat ballooning and collapsing as it sang. The light caught its eyes, twin points that refused to blink. My father inched forward, each step patient as if he had done this his whole life. The gig hovered, ready. The frog's throat swelled again. Then, with a sudden thrust, my father struck. The prongs drove through, and the sound cut off mid-call. He lifted it high, impaled and flailing, and grinned as though he had performed a magician's trick.

Two more followed, smaller but still heavy in his hands. He declared we had enough and handed one to me. Its legs dangled, slick and cold. I gripped them tight, though they slipped in my palm, and I felt a surge of pride for not dropping it. Only once did it slide free before I caught it again.

The strangest part of that night, the part that stays sharpest, came on the ride home. My father set the frogs in the back seat, propped upright, their legs forward and bodies slumped. In the glow from the dashboard, they looked like three little aliens seated for a journey to another world. Their glassy eyes stared into the dark as the car rattled down the road. The sight cracked something loose in me. I laughed until my ribs ached. For once, my laughter made him laugh, too. For a moment, we were just father and son, sharing something that did not sting.

At home, Mom fried the legs in cornmeal and lard. I hesitated, but the taste surprised me. They were good, with a crisp exterior and tender interior. Johnny sat horrified, clutching his biscuit, refusing to touch them. I wanted more. I wanted the night to stretch longer, with frogs and laughter and no hint of anger. But that was not how things worked in our house. Moments of light passed quickly, swallowed by darker days. Still, that night remains bright as a lantern in a mine, one of the few where I saw my father's face softened by joy.

Looking back now, I see how rare that night was, how unusual it was for my father to show me something that felt almost like pride instead of scorn. It is strange what sticks. I do not remember the details of school lessons from that week, or the chores I surely did, or what I dreamed about when I finally fell asleep. But I remember the frogs. I remember the smell of damp earth, the way the carbide lamp hissed like a breathing thing, the way my father's big hand felt on my shoulder when he steadied me. Fear lived in me, but so did a flicker of belonging.

The more I think about it, the more I realize that night with frogs was not just about food. It was about control. My father lived in a world where strength mattered, where a man proved himself by what he could kill or catch or break. To him, teaching me to gig frogs was not cruelty but preparation. He wanted me to know that if the world went hard, if money dried up and cupboards emptied, I could still find something to eat. It was his version of love, bent and twisted though it was. Love in our house came dressed as survival.

I have told people since then about frog gigging, and they smile as if it were a simple adventure, a boy learning to hunt with his father. I rarely tell them about fear, the smell, the cold slip of frog legs in my palm. They want the story of a country boy's fun, not the truth of standing half sick at the edge of a tank while a man you feared most in the world ordered you forward. I let them believe the lighter version. It is easier than explaining how pride and dread can mix until you cannot tell which one drives you.

Even now, when I see frogs along a riverbank, their throats pulsing, their calls rolling low across the water, I think of that night. I think of those alien eyes glowing in the headlamp, and I think of my father's laughter in the car. A part of me longs for it, not the frogs or the killing but the brief warmth of being let inside his circle. I suppose that is what children of rough fathers always want. Not approval exactly, but a moment of peace, a moment of being seen.

I sometimes wonder what my mother thought when we returned home with frogs. Did she see in his grin a glimmer of the man she once believed he was? Did she hope for more nights like that, where laughter replaced rage? Or did she only see the mess, the grease spattering on her stove, the smell of frying legs clinging to curtains? She never said. In our house, silence covered more than words ever did.

Johnny told me years later that he never forgot the sight of those legs twitching in the pan, how he swore off frog legs forever. For me, the taste never carried the same horror. I liked them, though I never sought them out again. What I held was not the flavor but the strangeness, the night of half fear and half belonging. The frogs became a marker in memory, a reminder that even storms can hold a minute of calm.

That memory, when I put it next to the others from those years, feels brighter because so many of the rest are dark. It is a strange thing when a boy treasures a night of killing frogs simply because it was not worse. After all, it was one of the few times my father laughed with me. But that is how memory works. It chooses what to keep, and it shapes itself into something you can carry.

It was a moment of education and of joy. It was not in letters or books, but in the language of survival and fear. I have not shied from the grotesque humor of it. Joy and education, for me, were rarely found together. I put it down here so it will not slip away.

CHAPTER 06

Wild Man

I woke on Saturday with a voice lodged in my skull, not gentle or thoughtful but grating, sharp as a rasp. It barked accusations I already knew were true.

"You! You'd be a fool to throw money away on some rundown flophouse for three or four weeks. How do you expect to scrape together more rent on the pitiful wages you make laying brick? You have a wife and child who come first!"

The voice was right. I had a wife and a daughter, though the wife had already slipped from my grasp. Three years of marriage, and she had filed for divorce…all I could see when I looked at my future was gray emptiness. I didn't have the language to say it was depression, but that's what it was. A fog that made every decision feel like groping in the dark. The papers weren't final, but the verdict was. Separation hung between us like a wall. Our daughter deserved better than a father still chasing illusions, but all I could see when I looked at my future was gray emptiness. It was not the twenty-five-mile drive that made me consider renting a bed in Deco's house near campus. It was the hunger for some different horizon, any horizon at all.

I had a girlfriend, too, unconventional and exciting, but she came with costs of her own. How would she fit into a house full of students? Still, the idea of living in a college town drew me like a magnet. I couldn't shake the pull, even though I knew it threatened to tangle everything further.

My marriage wasn't the only force pushing me toward change. My childhood had been unstable, brutal more often than not, especially after my father came home from prison. That brutality bled into every corner of my life, shaping the boy I was and the man I was becoming.

My younger brother John and I both dropped out of school before finishing. He made it to eleventh grade; I only staggered to ninth, and I was already two years older than my classmates after repeating the third and fifth grades. We moved so often that school never had time to take root. Our father, unpredictable and frequently violent, saw to that.

When he wasn't locked away, he was employed as a bricklayer. That was his trade, and it became mine, though he couldn't hold a job more than a few weeks at a stretch. The script was always the same: a foreman mentioned a crooked brick and then made a minor correction. Dad would brood, his anger simmering, then explode. Tools clattered to the ground. He'd storm over to his boss, red-faced and snarling. "You brainless son-of-a-bitch. I quit this sorry job!" Strangely, he treated each new job like a bar fight he hadn't yet started. The foreman never knew he was stepping into the ring until the bell rang.

If the foreman wasn't quick or cautious, my father's right fist followed, smashing into a jaw, dropping the man cold. Sometimes he kept hitting until the other workers pulled him off. And that was that. He'd be fired, blacklisted, and we'd move on to the next town, where the pattern repeated.

Jim Benson, my father, had been a wild boy before he became a wild man. He and four brothers grew up on a hardscrabble farm north of Dallas, near Aubrey, Texas. The family's clapboard house leaned tiredly against the dirt road. They drew water from a hand-dug well, cooked on a wood-burning stove, and trekked to an outhouse with a crescent moon cut in the door. Even the moon shape had been gnawed wider by a woodpecker hunting for insects. Inside, a battered Sears catalog waited, its glossy pages bound for a less noble use than shopping.

I hated visiting that farm. My uncles took delight in tormenting me, each in his own way. Uncle Clyde, the eldest, was the meanest. His face seemed carved into a permanent scowl. His pocketknife never left his hand, and he snapped it open and shut with a cruel kind of rhythm. One afternoon, I sat playing with jacks from an old cigar box when Clyde appeared in the doorway. His shadow stretched across the floorboards, dark as his grin.

"Hey Bobby, come sit on my lap. Your damn ears are too big. Ain't none of us Bensons got ears like that. Must be your bony mother whoring around while your daddy was off in the Army. I got something in my pocket to fix those ears."

I didn't understand "whoring around," but I understood his tone. My stomach knotted, my hands froze, and I looked for my mother or grandmother Ma. No one came. The only exits were the front door, which he had blocked, and the kitchen door behind me. My heart slammed in my chest. I dropped the jacks and bolted, my short legs pounding across the room. I wasn't fast, but in my mind, I thought if I pumped hard enough, my legs might sprout wheels.

I tore into the kitchen. Empty. Grandma wasn't there. Panic sent me further into the bedroom. Deserted, too. Trapped. The only refuge was under the iron-framed bed. I crawled as far back as I could, pressing myself against the wall. His boots followed, stopping just feet away.

"You little bastard, thought you'd get away, didn't you?"

He dropped to his knees, reaching under, his fingers scrabbling just short of my ankle. He pulled back, dug into his pocket, and drew out the knife. The blade winked in the dim light.

"Now ain't this a fine place to be? You might as well come out. I got you cornered. I'll just pick up this bed and get them ears off, quick."

Terror broke my throat open. "MOMMA!"

He stood, grinning with yellow teeth, and in one heave yanked the bed from the wall. The headboard screeched on the wood floor. He caught my ankle and dragged me out, limp like a rag. My lungs seized, my eyelids clamped shut. I waited for steel to bite bones, for hot blood to run down my face.

I hated Aubrey, Texas: its scorched cotton rows, its dirt roads, the August heat that clung like a fever. Clyde's knife seemed no different from the wire cutters farmers used to notch pigs' ears or the pliers they used to twist off turkeys' toes. Each mutilated creature carried its owner's signature. Come autumn, a single shot brought down a turkey, and the missing toes proved who claimed it. That was the logic of survival in those parts, cruelty branded as courtesy. It took me years to see that same cruelty at work in people, too, how a family could carve marks into its children and call it tradition.

But just as the blade caught the light, Grandma Glenn filled the doorway. "Clyde, let that boy go this instant!" Her voice cracked like a whip. "I warned you. Leave him be. You reckon you're a man, but I'll tan your hide if you don't stop this foolishness. He's nothing but a child, and he thinks you'll lop off his ears. Put that blade away and set him down."

I don't know if Uncle Clyde ever truly meant to cut my ears, but in that moment, with Grandma's stormy-gray eyes blazing, I believed it. She was a small woman with a back straight as a ruler, her anger sharpened by Irish fire. My brothers and I called her Ma, and when she raised her voice, even Clyde obeyed.

Another memory from that scorching summer of 1950 flickers like an old film reel. I was six and was sent back to my father's parents' place. Ma was hot fire but my grandfather, Big Johnny, was calm at her center. He was tall, broad-shouldered, and quiet. When he spoke, his voice was low and gentle, carrying more weight than a shout.

That August day boiled with heat. Big Johnny decided it was time to change the oil in his battered car. He'd jacked it up with a rusty bumper jack and crawled under to reach the plug. Sweat ran into his eyes, his hands slick with grime. I played nearby in the dust. Without warning, he went still. A heart attack hit him right under the car.

I remember Ma's scream when she saw him. She rushed out, trying to drag him clear. Her thin frame strained against his weight. She braced her knee against the car, sobbing, calling on God. The jack slipped, and the car slammed down across his lower body. I can still hear the metallic crunch of that undercarriage crushing Big Johnny's check, and my grandmother's wails echoing across the yard. What happened after that vision remains hazy, but her grief etched itself into me. I can still see her on her knees, crying over him, as if her anguish alone could call him back. That was the first time I understood that grief had no power, that it could shake the earth and still change nothing.

Years later, I would lie awake in Seguin, staring at the ceiling, weighing choices that seemed impossible. Should I move into Deco's house and the other students, at least for a few weeks? My brother had managed there for two years. If he too could work as a bricklayer and still go to college, maybe I would have a chance, even with nothing more than a GED I'd scraped together in the Army. The thought of college felt absurd but hope in me flickered.

Yet my obligations pulled me the other way: a wife halfway gone, a baby daughter who needed a steady father, a future that seemed like quicksand. I lay there torn, wondering if I was chasing something foolish or finally reaching for something worth holding.

Still, the lack of decision gnawed at me. Was I being reckless, selfish, even foolish? Sleep wouldn't come. The highchair. The cold water. The voice that shook the plates. Those memories never stayed buried long.

CHAPTER 07

Tubby

My grandparents' house was small but much newer than the house where my two elderly aunts lived. But the ways they lived had retained a strong influence from their own childhood in the 1800s. They had a Jersey milk cow that I learned to milk as a child. They had chickens and a big vegetable garden growing something all year long. They had a root cellar under the house, stocked with Mason jars of canned foods to last them through the winter. They had a corn crib half full of shucked corn cobs, and every year, they raised a pig and slaughtered it in the winter. If the wind was in the right direction, you could smell the stink wafting from the pig pen. Tonight was one of those nights.

Odors can transport you back in time and to childhood places. The sweet smell of butterscotch candy takes me back to my other grandmother, Grandma Glenn, and her wood-burning stove. It takes me back to burnt-sugar pies. They were incredibly wonderful. I think today, you'd call them butterscotch pies.

For example, the smell of hot sulfur transports me to Pasadena, or "Stinkadena" as some called it, another place we lived as children. When the wind was from the north, the scent didn't sneak up on you; it barreled in from the paper mill on the Houston Ship Channel. The paper mill emitted a pungent, nose-wrinkling cocktail that people described as a blend of rotten cabbage, spoiled eggs, and a hint of burnt wood. The stench originated from sulfur compounds released during the pulping process, which breaks down wood into pulp using harsh chemicals. It clings to the air and settles in your memory, a kind of industrial funk that's hard to forget once it gets in your nostrils.

There is a little town not too far from my grandparents' home, called Luling. Luling has many shallow oil wells that reek of sulfur gas. The smell of Pasadena and of Luling always moves my mind to thoughts of my first girlfriend, Shirley McGlothlin. Shirley was a bright, attractive fifteen-year-old girl with an infectious passion for life; her fiery red hair perfectly matched her spirited personality. Shirley's family had relatives in Luling, and during one visit, her parents, Dave and Mildred, invited me to join them. Though the details of the Luling trip have faded, my times with Shirley remain vivid. We were both brimming with the intensity of teenage hormones, unable to keep our hands off each other. We were careful to restrain ourselves in front of Shirley's father, but Mildred found our public displays of affection perfectly normal for kids our age. Perhaps watching us with amusement and indulgence brought back memories of her own youthful years.

This story is about me and my life. My life was shaped by both my genetic makeup and the events I lived through. My father played an oversized role. So, it is necessary to paint his picture as I came to know him.

James Albert Benson lived every day with a burning ambition to strike it rich, to break free from the chains of his failed life and answer to no one else on Earth but himself. His mind churned incessantly with feverish get-rich-quick schemes, each doomed to implode spectacularly. These disastrous endeavors invariably led to crushing debt, the cold confines of prison, or a devastating combination of both.

My father was a wild and uncontrollable child, harboring a disdain for most of the norms of civilization. Growing up in a dilapidated old farmhouse with peeling paint and sagging porches, where he and his four brothers were raised, there was a blatant disregard for others that permeated the air like the smell of damp wood. Serious crimes were committed with reckless abandon, showing no concern for neighbors' safety or the law's repercussions. Down the dirt road, not far from where my father and my uncle Clyde would lounge on the creaking front porch, a black family lived. My father and Clyde, gripping a rusty old single-shot .22 rifle with mischievous glee, would wait for the black children to pass on their way to school. As the children approached, gunfire erupted, the bullets kicking up dirt just in front of their feet, forcing them to stumble back, dance around, and sprint away in terror from the violence unleashed upon them. This was just the sort of twisted sense of humor that Clyde and my father relished, finding laughter in the fear they instilled.

Jim Benson disliked many things, but he loved horses, especially those belonging to others. At age 16, his passion for horses reached fever level, and his daydreams were filled with images of cowboys and the Old West, gun slingers and desperados, riding through dusty trails. He longed to be whisked back in time, galloping alongside notorious outlaws like Bonnie and Clyde, Al Capone, and especially Jesse James, who was his childhood hero and traveled by horse. The desire for his own horse consumed him. He was fixated on the Appaloosa gelding penned in a corral on the outskirts of Aubrey, picturing himself riding him into town to stage a bank heist.

One moonless night, he succumbed to his yearning and crept out to the man's corral. His heart pounded with adrenaline as he stole the horse. He rode it bareback under the starlit sky, feeling the wind rush through his hair as he galloped through dark open fields. When dawn approached, he led the horse into the woods, securing it to a tree, hidden from prying eyes.

When the sun broke over the horizon the next morning, casting a golden glow on the dewy grass, the owner of the ranch discovered his prized horse was missing. Panic gripped him as he recalled the local troublemaker, one of the Benson boys, and he wasted no time dialing the Denton County sheriff, hoping to resolve the matter swiftly. Tubby was my father's childhood nickname. The moniker stemmed from his cherubic baby roundness, owing to the way he had been nursed on rich Irish mother's breast milk. He was a pale-skinned baby with a penchant for loud, attention-seeking wails. A brief but thorough search of the shadowy woods behind the Benson farmhouse unveiled the tied-up pony, and with it, my father's journey into a lifetime of run-ins with the law began.

Sixteen-year-old James Benson, still known as Tubby to friends and family, stood nervously in front of the county judge, his hands shoved deep into his pockets. His imposing father, Big Johnny, and his quick-witted Irish mother, Ma, were summoned to the courtroom, their expressions a mix of worry and defiance, as Tubby, still a minor, faced the weight of the law. The judge, a stern man with a gravelly voice that echoed off the wooden walls, dominated the room with his words.

"Mr. and Mrs. Benson," he began, his gaze piercing like daggers, "I'm inclined to send your thieving son to prison as an adult. He is an unruly young man, a disgrace to the community of Aubrey, and should be a disgrace to you. All your brood, save your poor, unfortunate daughter, is not worth the sweet corn you feed them. Not one of your boys is going to amount to five dollars' worth of beggar-picked cotton. So, do either of you have anything to say before I decide what's to be done with this sorry juvenile horse thief?"

Big Johnny, with his weathered face resolutely set in silence, seemed to absorb the judge's words like a punch to the gut. However, my grandmother Glenn, fiery and determined, would not allow her son's fate to be dictated without a fight. Her voice rose above the tension in the room, rich with the cadence of her Irish heritage. "Don't you dare blame his horse thieving on me or my man! Tubby's had more belts stinging his butt than any kid you ever saw. We might be poor and not up to your livin', but we know how to raise an ornery kid. We don't spare the rod." Her fierce words hung in the air, challenging the authority of the judge while defending her family's honor.

"Well, I'm supposing you'll know what happens to young boys that I send to Huntsville. They get their due. It's like a meat market down there. Some deranged and tattooed convicts will start a jail-cell war over who's going to get him and take him for a slave wife. You don't want to think about what he'll have to do to stay alive. It ain't pretty."

Ma looked confused. "I don't know about what you say, or what you're trying to tell me. That boy's got Choctaw blood in him, and that's why he's like he is. All them Indians is mean, like my son. My boy's tough."

Throughout this court proceeding, my future father had been silent. But now the judge addressed him directly.

Well, boy, here's what I'm going to do. Either you're going to enlist in the Army and let them teach you some right and wrong, or I'm going to set a date for a trial, and you will be found guilty. In Texas, stealing a horse is a third-degree felony. This carries a prison sentence of up to 10 years and a fine of $10,000."

As you might expect, my father opted for the Army, a choice that thrilled him to his core. He envisioned himself in crisp, olive-green fatigues, marching in perfect formation, charging into battle, and experiencing the adrenaline of war. However, reality struck him hard; at just 16 years old, he was too young to enlist in the military. With his head drooping in disappointment, he muttered, "I want the Army, but they won't let me in," his voice barely above a whisper.

The judge, a stern figure with glasses perched low on his nose, gestured for Big Johnny and Ma to step closer to the bench. In a subdued tone, he explained the gravity of the situation: their son would need to lie about his age. "The recruiter will see right through the lie," he said, "but if he claims he's 18, they'll take him alright. They need fodder in case of a war."

My dad's eyes lit up with excitement at the prospect. The war in Europe was already raging, and the thought of fighting the Germans filled James Benson with a sense of purpose he had never known. The judge granted him a month to enlist, and without hesitation, he made it happen.

Arriving at Fort Sam Houston in San Antonio, Texas, he puffed out his chest in pride, clutching a rifle that felt heavy yet empowering in his hands, learning the precise steps of marching under the hot Texas sun. His dreams of war danced in his mind like a flickering flame. His commanding officer, impressed by his fervor, selected him for specialized training.

My father was sent to Minnesota, where he joined the 1st Ranger Battalion commanded by Major William O. Darby. There, he mastered the art of skiing through thick drifts of snow while wearing a heavy backpack loaded with gear and weapons, each lesson a reminder of the challenges ahead. He practiced parachuting from transport planes wearing full battle gear. He honed his skills, learning the terrifying techniques of dispatching enemies in hand-to-hand combat with his bayonet. They taught him the art of creeping up behind unsuspecting German soldiers and grimly detaching their heads with a piano wire around their throats, like slicing through a hock of ham. All, preparing him for the brutal realities of war that he so eagerly anticipated.

However, he was a distraught child, and the bloody and brutal realities of war altered him irreversibly. The constant presence of mutilated bodies and death, along with being shot three times himself, pushed him beyond the brink from near insanity to fully spiraling into madness. He endured the perilous landing on Normandy

Beach, scaled the towering cliffs into France, fought fiercely in the Battle of the Bulge, earned the prestigious Silver Star for his valor, and returned to Texas with a mind shattered beyond repair. As you will see, his life profoundly overshadowed and twisted my own.

CHAPTER 08

The Turkey Massacre

The Hills. A place that captured my imagination as a child. In my mind, these hills formed one point on a big mental triangle that stretched from the fishing camp on the Guadalupe River, to the Capoté Knob, and then to the old ranch house near Leesville, where bad things happened. With a ruler, I drew out the triangle with a pencil. It was very close to an equilateral triangle, with the first leg almost aligned east-west. Leg one was 5.07 miles, turning south south-east for 6.73 miles, and then turning north north-east back to the river camp, a total of 18.25 miles.

Of all the places we lived when I was a child, the old ranch house was the most dismal. Maybe it should be characterized as a shack. Dad had taken a job as a cowboy on a ranch owned by Zillman Davis and his family. The place was a little east of Leesville, and on a dirt road to nowhere. There was no electricity, no well for water, and the roof leaked when it rained. We were also instructed to avoid the broken boards that created gaping holes in the bedroom floor. There were cracks in the walls, and some of the glass panes in the windows were broken.

Unlike our previous home, there were no roaches in the house; however, we did have termites, spiders, and scorpions, but not roaches. No creatures other than rats and raccoons had occupied the "dwelling" for years, so there were no crumbs for roaches to eat.

One of the first repairs that Dad did when we moved to Leesville was to patch a rusty hole in the bottom of the house's cistern with black tar. The cistern was rigged to collect rainwater from the tin roof. I couldn't believe that we were going to drink our water out of this nasty thing at our most recent place.

While Dad was repairing the hole, he took a break to fetch a room-temperature Lone Star beer from the trunk of our dilapidated car. I climbed up his ladder and peered down into the cistern. All sorts of debris had accumulated on the bottom of that great tank. It was mainly leaves and sticks, but I saw something that looked like a horse bridle and one faded high-heeled shoe. When Dad got back from swigging his beer, I asked him about the trash I had seen in the cistern.

"Daddy, I saw a shoe in the tank. Are you going to clean it out?"

"Bobby, you little bastard, how did you see a shoe? Did you climb up on that ladder?

This was not good. I put my chin on my chest and didn't answer his question.

"You better, damnit, look up at me. Were you up on this ladder?"

I knew from experience where this was going, "Yes, Sir. I thought it was okay, since you wasn't working just now. I wanted to see what it looked like in there."

"Come the hell over here right now. I guess you don't learn too easy. You gotta another lesson coming."

He took off his belt, grabbed my arm, and began administering my expected punishment. He began lashing my bottom with his leather strap, over and over. I'd learned from repeated experience that I would do better to start crying. He would never stop until I was bawling like a baby. It hurt so badly that I tried to run away, but since he was holding my arm so tightly, all I could do was run in circles. We performed a hideous merry-go-round dance, and then he finally stopped.

"Now I'm going to answer your goddamned question. If I can reach that shoe with a stick, I'll pull it up. If I can't, then it will stay down at the bottom and be under the water with all that other shit that's in there. It ain't going to hurt you if our drinking water is a little bit brown. People's been drinking out of rivers full of trash for hundreds of years, and they didn't mind. You're no better than them."

I nodded my head but could only answer with sniffles.

"And one last thing. Don't you dare tell your mother what you saw in that tank. Do you understand me?" I nodded.

I don't remember exactly how old I was, but I'm pretty sure I was twelve. I was not aware of the details of my father's activities, but I've come to understand that they were mostly illegal. This move to Leesville, an old, rundown shack-of-a-house, could have been the fourth place we'd tried to live that year. My father was on the run from the law. He had been in prison more than once by this time, and neither my brother John nor I knew about this. Mom convinced us, my brother and I, that our father was in the "hospital" when he was picking cotton on a prison farm somewhere around Houston. I can't count the times that John and I were jerked out of school and moved to a different school during my childhood. In the 1950s, unlike many schools today, an underperforming child could "fail" (be made to repeat a year of school), and that happened to me twice. I failed the third grade and was required to repeat it. Again, I failed the fifth grade and had to repeat the year. My brother John was born two years after me, but owing to my repeats, in one school year we were both in the same grade.

John and I tried to stay as far from Dad as possible. His hair-trigger violent episodes were unpredictable and frequent. We kept our mouths shut and our heads down. We were both enrolled in the Nixon School District, the elementary school, and we rode the bus five days a week. The bus picked us up a little after six in the morning, and it was a 25-mile ride, circling all the dirt roads in the northeast part of the county. It circled through the Quien Sabe Ranch, home of the small German community near the Capoté Hills. We were the first kids to get on the early morning bus and the last to get off in the late afternoons. One of the German-speaking boys from the Quien Sabe Ranch would always sit by me, and he tried to teach me my first German words. Klein may have been my first. I learned a few phrases like kleines Haus (small house) and mein kleiner Bruder (my little brother). That was John. Klein is one of those versatile words that threads its way into everything from poetry to physics. Another kid, a little older than me, who rode our bus and lived nearby was Wilbon Davis. Wilbon's family owned the ranch where my father was employed, and I believe that his family also owned the old house we were occupying. Wilbon's father, Zillman, loaned our father a horse. Dad rode that horse to work at the ranch every day. Since I was younger than Wilbon and one of the peons on their ranch, I didn't know him well, but he later played an oversized role in my life.

Owing to Dad's strictness and downright brutality, John and I stayed away from the house as much as we could. If we weren't going to school or being at school, we roamed the bleak countryside, playing and exploring, and daydreaming about what we would do when we grew up. Once we discovered a Red-tailed Hawk's nest, we regrettably stole the downy-feathered babies from the nest and tried to feed them rabbits we shot with a BeeBee gun. They died within a few days. We were trudging across a plowed field and delightedly discovered a cache of arrowheads exposed on the sandy surface.

Sometimes, when Mom and Dad went to Dallas for a few days to visit his family, at least "visit" is what Dad said. They would drop us off at Mimmy and Grandpa's house, where I was allowed to listen to Grandpa's radio when he wasn't listening to the news. I loved that radio. I was mystified by it. How could people's voices travel through the air from so far away and come right to Grandpa's radio? How was that possible? It was magical, spiritual even.

It told me about a world far beyond Dewville and Leesville. Of course, my favorite radio program was Bobby Benson (my namesake) and the B-Bar-B Riders, a popular children's Western radio program that initially aired in 1932, then returned in a more enduring form in the 1950s. It followed young Bobby Benson, who inherited the B-Bar-B Ranch and led a cast of colorful cowhands through adventures across the American West. I remember listening to the election returns in 1952, when Dwight D. Eisenhower became president. I remember hearing Albert Einstein's high-pitched voice, which was accompanied by a squeaky tone and a heavy German accent. I heard words like genius, scientist, nuclear bombs, and the weirdness of special and general relativity, none of which I understood the tiniest fraction.

But John and I talked about these things. We concocted our own ideas about things like the speed of light being constant, the possibility of length contractions, and the seemingly impossible aspects of time dilations. The distance between my father's leather belt and the science of relativity was "infinity" itself. But speculating about these things gave us hope that someday we would be free.

After my dad was discharged from the Army after the war, he was entitled to government money to go to school and learn a trade. It was called the GI Bill or, more officially, the Servicemen's Readjustment Act of 1944. He had tried to learn about radio technology, but that topic did not fit well with his deranged and primarily criminal mind. I stumbled upon an old, worn book on radio electronics that he had discarded, and I was mesmerized by the diagrams of vacuum tubes, resistors, condensers, and other components. Like Einstein's voice, these pages magically connected me through the ether to another world. It was pure escapism, a coping crutch for a brutalized child.

I can't leave the topic of Leesville without relating another event that occurred during our time, which was marked by violence and brutality. Dad was always looking for a way to make it big. This time, he was going into the turkey business. There were lots of people around that part of Texas who had been snookered into raising "fodder" (by that I mean baby chicks and turkeys) for the big packing houses across the nation. There was a packing plant right in Nixon, where John and I were in school.

It is easier for me to talk about the things that Dad did to animals than the things he did to me. Fragmented flashbacks represent my memories of his physical violence toward me and my brother; partial scenes that don't always connect in a logical sequence. I'm not going to dwell indefinitely on this subject, but I will say enough to show the sort of man he was. In retrospect, I don't blame him for my treatment. He was a product of his own raising and his experiences in the War. His philosophy of child rearing is summed up in the old saw he used, "If it don't kill you, it makes you stronger." There's even a popular song, which was perhaps only popular in Texas, called "A Boy Named Sue," that glorifies this theme.

So here is another character-building incident that looms from our life at Leesville. The old, rotted house in which we lived had a wooden step at the front door. One day, the step broke. Dad decided to replace it with bricks. The bricks were left from the top of a hand-dug well, a man-sized hole still in the ground, that was used before the house was upgraded to a cistern for catching rainwater.

"Hey, Bobby, Johnny, you two come over here."

We didn't know what was up, so we were immediately scared. Had we done something wrong? Was he about to pull off his belt and whip us? Reluctantly, we complied with his order.

"Okay, I've got a job for you two. All you two do around here is go off in the woods, and I don't see y'all 'til suppertime. I want you to go 'round next to that old well and load up those bricks and stack them right here by the steps."

Neither of us liked the idea of fooling around the old well, because it was run through with Bermuda grass stolons (I didn't know that word back then), and I'd seen scorpions and black widow spiders up between the bricks. My grandmother told me that black widows could kill me, but what Dad might do to us might be worse. We headed to the brick pile.

As we arrived back with the first load, Dad unloaded two heavy bags from his car. One was a bag of mortar mix, and the other was a bag of sand. He dragged up a washtub and fetched a bucket of brown water from the cistern. He then started mixing up what he called mud. Had you been there, you might have noticed a rusted-out truck parked behind the house and wondered why Dad hadn't used it to go to Nixon and carry back those heavy bags to make the mud. He sometimes did use it to drive around the ranch, but he never took it to town. It didn't have license plates, and it didn't have floorboards for your feet. Those had long rotted out. And worst of all, the gas tank had rusted through so that most folks would have hauled it to the junk yard. But not our dad. He had rigged up a tank from a chicken brooder and routed a copper pipe from the tank to the top of the old truck's carburetor. He'd inserted a cock valve in the copper line and would feed gasoline from the brooder tank directly into the carburetor. He could start the truck and control its speed by turning the cock valve and pouring more or less gas down the throat. People used to be more innovative than they are today.

My grandfather, Louis (Grandpa), was very innovative in a nineteenth-century way. After building their retirement home on Mimmy's inherited property at Dewville, they needed a water well. He hired a well digger, and they had running water. The water had an awful taste - a taste of rusty iron. He designed his filter system, a rectangular tank with three partitions. Between each partition was a six-inch-wide section through which well water must pass to get to the house. The first filter was sand from his back pasture, and the second filter was charcoal he'd made by burning mesquite wood and then smothering the fire with sand. This did render the water more drinkable, and Gandpa was proud of his ingenuity. But an unforeseen problem arose. The mosquitoes had open access to the water passing through the filter, and soon the sections were bubbling with mosquito larvae. His solution was to go fishing. He returned with a bucket of minnows and small sunfish. He dumped these into the filter tank to eat the mosquito larvae, and for years, we kids drank well water there without knowing it was full of fish shit. But we lived.

"Hey, you two. Get the hell over here with those bricks. I want to get these steps done by dark."

I responded, "But the wheelbarrow has a flat tire, and it's hard to push with all these bricks."

"Hell no. Just push it flat. Make Johnny help you. You take one handle and let Johnny take the other one. I'll fix that tire tomorrow."

I knew better than to argue about it. But I didn't let Johnny help me. He was smaller than me, and we had already tried pushing as Dad demanded. I just gritted my teeth and put all my strength into the push. I could move it, but not too fast. I put Johnny to herding the half-grown baby turkeys out of my way. They were loose and pecking around everywhere, always getting in the way, no matter where you stepped. Dad told us that, "A turkey is the dumbest animal in the world. If it rains, they open their mouths and look up at the sky and drown themselves."

I didn't know how many loads of brick we needed to dig from the brick pile and haul to the construction site, but we made the loads smaller to improve the rolling. We planned to make as many trips as necessary.

Dad was making headway on building the steps, but the turkeys hampered his progress. They kept jumping up on the bricks, adding enough weight to screw up the alignment by pushing a corner down into the still-soft mud. Dad hated crooked bricks. He was getting increasingly frustrated and cursing louder.

"Johnny, get your little ass over here and help me keep these goddamned turkeys off my bricks. Let Bobby get the bricks by himself."

Johnny started chasing the poults around, but all that did was stir them up and make them wilder. His efforts were making the problem worse, and Dad's fury escalated. His Irish face, inherited from Grandma Glenn, was turning red, and his cursing got louder and began including short, striking words we'd never heard before, and we didn't know their meaning.

Finally, Dad's intense anger reached an inflection point, and he exploded in rage. While on his knees, he started violently slinging the sharp-edged trowel in great arcs, chopping through the necks of baby turkeys and severing whole bodies. After massacring five or six of the closest ones, he got on his feet and began chasing down and executing other birds. Before it was over, the yard was littered with blood and dead turkeys. Mom came running out of the house to see what was causing the commotion in the yard. She was grabbing Dad's arm and trying to get him to stop, but she only weighed about 100 pounds and couldn't stop him. John and I were horrified. It occurred to me that, when he finished with the Turkeys, he might go for Johnny and me. This was the size of rage I'd seen only when he was in a fist fight with other men. Or maybe it was like what he did in the war, cutting the throats of German soldiers with knives or piano strings.

Finally, he settled down, dropped his trowel on the bloody ground, shook our mother away, and headed for the trunk of his car to get another beer.

I've not said much about our mother, but you should know that she was unique among women. Her marriage to Jim Benson was littered with turmoil and long periods of stress and poverty as he was in prison. How she endured her shattered life amazes me. I think when she fell for him as a young soldier, she committed herself to the marriage unconditionally. "...to have and to hold from this day forward, for better, for worse, for richer, for poorer, in sickness and in health, until death do us part." She lived those words.

CHAPTER 09

Fresh Meat

I knocked on the door. No answer. I knocked on the door a second time, more insistently, then a third. I was just about to hit it again when Deco's voice growled through the wood. Somewhere between a yell and a grumble, I heard him drag out the words, "Yeah---Who the hell is it?" His voice had the scratchy edge of someone pulled out of a deep sleep, as if every word scraped out angrily from the back of his throat. "Don't you know it's Sunday?" he went on, incredulous. "I can't believe this. Nobody's supposed to be waking me this early on a Sunday." He paused, as if waiting for some acknowledgment or a show of mercy. Instead, he continued to vent. "Go the hell away," he shouted. This time it was louder, as if he were using the last of his patience and a good chunk of his lung power. "

I hesitated for a moment, unsure of what to do. I stood in front of the door, debating whether to wait a bit longer or turn around and head back home. After all, it was ten o'clock in the morning, and the town was already wide awake around me. People were out mowing their lawns, walking their dogs, or just enjoying a little sunshine and fresh air. Surely, I thought, most of the world was on its feet by now. It wasn't as if I had barged in at sunrise. I glanced back at my car, wondering if I should swallow my pride and leave. I was about to call some parting words through the door, something like an apology or a sarcastic "See you later," when I heard a shuffle from inside. Then there was a thud, like someone knocking over a stack of magazines or maybe tripping over a stray shoe in the hallway.

"All right, I'll be there in a couple of minutes."

I pulled open the screen door, its hinges creaking softly, and carefully cracked open the solid wooden door behind it. The door opener was not the typical round knob but a sleek, elongated handle that required a firm downward push to release the latch. This kind of door opener was, if not unique, certainly a rare sight in the United States. Its design was reminiscent of those I'd first encountered in Germany during my time in the Army, a slight nod to European craftsmanship and a subtle reminder of my travels abroad.

I joined the military just after I turned 17. I may have needed my parents' consent to enlist at that age, but I don't remember exactly. After completing basic training,

I was dispatched to Fort Lewis, Washington, for a time before being sent to Germany. My experience in Germany was limited, as we were engaged in an extensive training exercise. My battalion spent most of our days in the field, enduring the biting winter chill of the Black Forest while living in cramped pup tents. On rare occasions when we were granted permission to visit the nearby villages, we were struck by small details, such as how the house doors opened with handles rather than the round knobs we were accustomed to in America. It may seem like a trivial observation, but it gave rise to a wistful phrase among us lonely soldiers: "I want to go home to the land of the round doorknobs."

Deco finally came to the door, his hair tousled and eyes puffy with fatigue. He squinted at me and exclaimed, "Oh shit, it's you. I didn't think you were coming in the middle of the night. Come in, and I'll make some coffee."

His disheveled appearance hinted at a long night with very little sleep. I trailed behind him into the kitchen, where the chaos of his late-night activities was evident. The sink overflowed with dirty dishes, while empty beer bottles cluttered the table. A precarious stack of textbooks and loose paper leaned haphazardly from a chair, threatening to spill onto the floor. Deco dug out an old percolator buried under a pile of plates in the sink, rinsing it hastily under the running water from the faucet.

"So, you decided to take the room, did you?"

"Yeah, for at least two or three weeks," I said, sliding two crisp twenty-dollar bills across the scratched Formica table. Sunlight filtered through grimy blinds, painting the torn vinyl tabletop in bands of gold and dust. Deco dumped a scoop of medium-roast grounds into a glass-knobbed coffee pot, the rich aroma blooming in the cramped kitchen.

"I brought some clothes and odds and ends," I continued, unzipping my bag with a soft whump. "I'm guessing you won't mind if my girlfriend visits sometimes. She might even crash here once or twice."

Deco shrugged, flicking on the burner beneath the pot. Women are always welcome. It's your room, so feel free to lock the door if you want. Please note that we keep odd hours and make noise. Folks wander through here all the time, especially on weekends. Tonight's our weekly poker game. You're required to play," he added with a crooked grin. He squinted at me over the rising steam. "We need fresh meat. You play poker, don't you?" His smile made me uneasy, as if I were a lamb at slaughter; a bricklayer with calloused hands and no high-school education against a flock of smooth-talking college kids.

I leaned back, brushing dust off my jeans. "What do y'all play, draw, or stud?"

Deco laughed, setting the coffee pot on a chipped coaster. "No, man, Texas hold 'em. Ever tried it?"

I had folded and bluffed plenty of hold 'em hands and won my share of pots. I kept that information to myself. In Texas, poker games for money were illegal. Illegal games could be police traps or setups for robberies. Winning might be easy, but safely getting out of town with your winnings was scary. A low-stakes student-house game sounded tame enough.

"I've heard of it, but I've only played it once or twice. I think I remember the rules. Count me in."

"Cool," he said, clapping me on the shoulder. "We'll save you a seat."

Steam curled from my mug after Deco shuffled off to crash for the morning. I wandered down the narrow hall, half-expecting footsteps behind me, then found the door to my room, a single bed, a metal desk, and a flickering bulb overhead. I dropped my bag, sat on the edge of the bed for a moment, and decided to stretch my legs on campus.

Outside, the air smelled of wet limestone and freshly mown grass. I found a pay phone and dialed up my brother John, the line crackling like static over the San Marcos hills. I invited him to lunch.

San Marcos, Texas, home to Southwest Texas State University, was known as a party town. The campus hugged the ancient Balcones fault line that divided the Texas Hill Country from the Blackland Prairie to the east. I climbed a hill and watched the clear water flowing in the San Marcos River, which was icy cold and so pure it looked like liquid glass. Everywhere I looked, students lounged under live oaks, the sun catching hair of every shade from honey blonde to deep mahogany. It was a beautiful campus, and precisely the kind of place I wanted to be and needed to be.

I met my brother at a sunlit café on South Guadalupe Street, the one with the chipped-red awning and sticky vinyl booths. We both ordered BLT sandwiches on toasted sourdough, with crispy bacon, iceberg lettuce that snapped under our teeth, and thick tomato slices that dripped onto the table. Between bites, he told me about his day, and I listened, watching him push lettuce around with his fork.

After lunch, we crossed the quad to the Science Building, where John took most of his classes. The façade was dull brick and narrow windows; inside, a chrome-lined elevator whisked us to the first floor. Physics labs occupied that level. There were long, fluorescent-lit halls lined with demonstration stations. One display, a cloud chamber, caught my eye immediately: a glass-topped box fogged from below, wisps of alcohol vapor swirling as invisible subatomic particles etched whitish streaks across the cloud. Tiny, ghostly lines darted and vanished. I watched the tracks, and I was spellbound. I could have stayed there all day.

My brother tugged at my sleeve and led me up the stairs to the second floor, where biology claimed the corridor. The air smelled thick of formaldehyde and bleach. Behind a long glass case, twenty-four jars stood at attention on a narrow shelf, each containing a pickled human embryo or fetus in clear liquid. Each dead body ranged from a translucent, pea-sized cluster to a fully formed infant curled in fetal position. I felt nauseous as I gazed at the last example: a newborn, eyes closed, fingers curled around its thumb. I realized I was staring too long; my brother gently pried my elbow. "Come on," he whispered, "let's go back to the cloud chamber." But the line of tiny, lifeless bodies had already burned itself into my mind.

That evening, I found myself at a battered poker table in my new, cramped living room. Only four of us sat there: Beside me, Deco, Jeff, and Paul (my new roommates) were there. A single bare bulb swung overhead, casting long shadows on the kitchen table. I sensed they'd pegged me as an easy mark, college kids versus a dumbass bricklayer. I'd half-expected marked cards or sleight of hand. Instead, Paul reached into his jeans pocket, pulled out a pristine, cellophane-wrapped deck, and with a crisp tear of plastic announced, "Okay, boys, let's get started."

He explained the rules in a steady voice: Texas Hold 'em, quarter-dollar small blind, fifty-cent big blind, pot-limit after the flop, dealer button moving left. I nodded, forcing confidence. Deco pushed a quarter out; Jeff tossed in fifty cents. When Paul slid two cards face down to each of us, I peeked first: Queen of Hearts, Five of Spades, nothing to write home about. I slid fifty cents forward and tried not to grimace. Deco and Jeff called.

Paul burned a card, then laid out the flop: King of Clubs, Five of Diamonds, Two of Hearts. My lone Five paired, but it wasn't much. Deco's voice cut through the lamplight: "Two bucks." Jeff matched him. I hesitated, then folded, folding in memories of glass jars and the faint echo of vapor trails. The pot swelled as the turn and river fell; by the time Deco revealed three Kings in his hand, he raked in twelve dollars, more than a week's rent.

For the next half-hour, I folded round after round, studying their tells: Deco's quick blink whenever he bluffed, Jeff's shifting foot under the table, Paul's exhale before a big bet. My stack dwindled. Finally, I caught a small pot, just enough to keep me in the game. The clock on the wall ticked past ten; my pulse raced at the thought of bluffing against seasoned card players. Better to wait for a real hand.

It came on the next deal. I peeked at eight of Diamonds and nine of Diamonds, heart in my throat. The flop fell: Ten of Diamonds, Four of Hearts, Jack of Clubs, a gut-shot straight, plus the spark of diamonds. Deco, perched on the edge of his chair with the most enormous pile of chips, tapped the table. I just called to draw him in. Jeff and Paul piled in more chips, too; the pot climbed toward eighteen dollars.

The turn card was a flash of red, the Queen of Diamonds. Now I held four diamonds in a row, a straight flush. My breath caught. One river card stood between me and a straight flush. Deco slid a twenty-dollar chip forward, Jeff folded with a shake of his head, and Paul glanced at me before bowing out. The pot swelled to nearly sixty dollars in the center. I toyed with my last chips, then raised, sending a clear message: confidence. Deco's eyes narrowed; he hesitated but called.

The river card hovered in the air as Paul burned the top card and slid the final card face-up: Two of Diamonds. My chest sank. I'd missed the straight flush. All I could do was watch as Deco dragged the vast pile of chips to his belly. We packed up as the fluorescent bulb flickered and went out. Outside, Deco lit a joint on the front steps; the night air carried the tang of weed and distant engines. He exhaled a smoke ring, then looked me in the eye.

"Hey Benson, I'm sorry you lost all that money," he said, voice half-laughing, "you're smarter than I thought, though. Maybe you should try college."

I shrugged, wincing as the memory of glass jars flickered behind my eyes. "Maybe so," I said, looking out into the warm night. "But I've only got an eighth-grade education and a GED from the Army. I don't think they'll let me in."

He blew another ring into the dark and grinned. "Man, you don't get it. If your pulse rate, your body temperature, and your IQ add up to a hundred, they'll take you." I pulled open the Beetle's trunk, dropped my bag inside, and under that dim streetlight, all I could do was laugh.

CHAPTER 10

Purgatory

After the poker game, I lay awake in the unfamiliar room, the ceiling fan cast shadows that danced like ghosts across the walls. The bed felt foreign beneath me, its sheets still crisp and smelling of detergent. The muffled laughter from the next dorm room seeped through the thin walls, reminding me that I was surrounded by people whose names I barely knew. My mind spun like a roulette wheel, fixated on the hope that getting into college might be the key to transforming my life.

The summer term II was drawing to a close, and according to Deco, most students were shelving their social activities to hunker down with their textbooks in preparation for final exams. I imagined them, faces illuminated by desk lamps, scribbling notes and flipping through pages. Deco had a way of teasing me about the supposedly daunting entrance requirements of SWT, but beneath his humor, I sensed a kernel of truth. Surely, there were standards for what incoming students should have mastered in high school.

Could I tackle college courses that demanded a solid grasp of algebra, history, English, and other subjects that my peers had likely aced? My memories of the equations from those few months of algebra I'd managed before dropping out of ninth grade were as hazy as fog on a winter morning. The symbols and numbers felt distant, like a language I had once spoken but could no longer understand.

All this uncertainty gnawed at me like a persistent itch I couldn't scratch. College wasn't free, and the thought of where I'd find the money for tuition, fees, and books kept me awake at night. The looming question of how I'd manage to feed myself and pay rent weighed heavily on my mind. Most of my clothes, worn and frayed at the edges, were far from the fashionable styles sported by other college students. But what plagued me most was my thick, low-class Texas drawl, a constant source of embarrassment. I could already imagine the whispers and stifled laughs echoing in the classroom, "Who is this redneck sitting among us?" Could my desperation to break free from my blue-collar roots propel me through all these hurdles? I was determined to try, but the path ahead was laden with more obstacles than I had foreseen.

The next morning, at the first light of dawn, I dragged myself out of bed and poured a bowl of Wheaties from the box I had brought over from my small apartment in Seguin. Feeling restless, I decided to explore the town and get to know my new surroundings. That early, the campus was eerily quiet, with most students still buried under their covers.

This morning was going to be pivotal. In the night, I had made up my mind. I was going to see if I could enroll for the fall semester. The first and scary step was to check in with my boss at the construction site and resign. He was not going to like my decision. I would make sure to tell Otto first, and assure him that we would still be friends, and that I was sorry that I'd not be there to help him down from the scaffolding at the end of these hot days. I knew he would understand.

I set off downhill, following the road that led to Purgatory Creek. It seemed like the perfect spot for some solitary reflection. Otto Galvan had once described Purgatory Creek as a wild yet beautiful place. Since he lived in New Braunfels, just a short drive down the Interstate from San Marcos, his knowledge of this part of Texas was unmatched. The creek was typically a dry bed, but when thunderstorms roared through the hills to the west, it could transform into a raging torrent, striking fear into the hearts of those who lived nearby.

The name "purgatory" echoed back to my memories of Catholic school, and I pondered how the creek had earned its name. The story, as Otto told it, dated back to the first settlers who stumbled upon this creek. Exhausted, hungry, and desperate for water, they discovered a large spring. But their relief was short-lived. The spring-fed pool was hemmed in by a swamp teeming with clouds of relentless mosquitoes, and fresh signs indicated a Mountain Lion had recently hunted a deer nearby. One weary settler exclaimed, "This is hell! No, it's even worse than hell." Another chimed in, "It's purgatory!" And with that, the name stuck. If either of those biblical realms truly exists, I'd argue purgatory is the lesser of the two evils.

The creek bed yawned before me, with its cracked clay floor flecked with pebbles and sunbaked twigs. I settled onto a broad, pale limestone ledge, the stone cool and smooth against my palms. A soft breeze stirred through the scrub oak and yaupon beyond, rattling dry leaves, yet the birdcalls remained sharp, bright Carolina wrens trilling nearby and the insistent "jay-jay!" of Blue Jays echoing up the hollow. Beneath it all, a low, constant roar drifted from Interstate 35, the highway slicing the landscape between San Antonio and Dallas–Fort Worth, tracing the Balcones uplift like the edge of a giant sipper attached to a limestone winter coat.

Up the creek, a pair of Blue Jays flitted between cedar boughs. Their raucous cries carried me back to Aubrey, Texas, and to a summer when my grandfather, Big Johnny, lay beneath his battered old car. I heard his wrench fall by his side, and Blue Jays were calling that day, too. It is funny how two unrelated things can fuse concrete and timeless memories. The only good thing about Aubrey was the birds. That's where I first learned to love them. Unlike people weighed down by their unique consciousness, birds were free and unconcerned as they moved through their lives. Unlike humans, who could do unspeakable things, like dragging their kin from under beds to cut off ears, birds lived in the moment. The past and the future didn't matter. I closed my eyes, felt the limestone's grain, breathed the dry heat and dust, and let myself drift toward the consequences of my decision. College might be my only chance for a different life.

Back at the weathered clapboard house on North Street, the paint peeling at the corners, I found Deco at the kitchen table. Steam curled up from a dented aluminum pot on the stove, carrying the sweet roast of fresh coffee. He looked up from a dog-eared newspaper, rubbed his jaw, stubble catching the morning light, and slid me a chipped mug.

I poured the liquid, which sputtered before settling into a warm, dark mirror. "I've been thinking," I said, my voice low. "I want to try college. But I don't know when to start or what steps to take."

He leaned back, folding his arms over a faded chambray shirt. "First thing: fill out an application, get your GED scores, and take them to the registrar. But before that, you need to see a counselor, hand in your applications, map out your courses, and stuff like that. You can't just walk in, though; you need an appointment."

I frowned. "Where do I go to see a counselor?"

Deco's eyebrows shot up. "Old Main, I think. It's the oldest building on campus. It's the building with the red tile roof and steeple towers at each corner. You can't miss it. If the counselors' offices aren't there, someone in Old Main will point you in the right direction."

Even considering the money I'd lost in last night's poker game, a good thing about quitting my job was that I'd been paid on Friday, so there was no debt owed in either direction. I decided to keep my Seguin apartment until I knew whether they'd accept me. Quitting was easier than I'd expected. The boss was gracious about it. Maybe he had anticipated my leaving. All he said was. "Well. Good luck, old man."

I finished my black coffee and set out for Old Main, clipboard in hand. Deco was right, it was easy to see. I asked the secretary where I could find a counselor. She peered over her glasses, tapped a schedule, and said their offices were temporarily down the hill in the new Evans Library Building. The fluorescent hallway smelled of fresh paint, and I followed her pointing finger to a door marked "Counseling Services."

Inside, the carpet muffled my footsteps. A stout man sat behind a desk piled high with manila folders. His pin-striped shirt was streaked at the cuffs, his suspenders stretched over a generous midsection. The collar gaped open, and a crooked tie dangled below a rumpled chin. On the wall hung a gilded frame holding a faded photo of two small children and a woman whose smile looked patient. He didn't glance up as I closed the door.

"Mr. Staples?" I cleared my throat. "I'm here about…counseling. College admissions."

Slowly, he lifted his head, eyes narrowing as if sniffing a sour scent. He reached into a shirt pocket, withdrew a frayed handkerchief, and blew his nose with a squelch. "Who the hell are you?"

I drew in a breath, balancing my backpack on one shoulder. "Robert Benson. A couple of students told me I might still qualify for the fall semester." My voice felt small in the cluttered room.

He leaned back, chair creaking, and studied me as if I were a cockroach on the floor. "You don't look like a student, and you sure as hell don't sound like one. Did you fill out an application last spring, like everyone else?"

"No, sir. I…just got the idea this weekend. They said Army vets sometimes get special consideration."

He snorted, rubbing at his nostrils. "So, you're military, huh? When were you discharged?"

I set my jaw. "The day after Christmas, 1962."

He rubbed his eyes, then sprayed a sneeze into the handkerchief. "That means you've only got about two and a half years of GI benefits left. Not nearly enough time to crank out a degree, and it won't pay much."

My cheeks burned. "I know time is tight. But if you let me in, I'll pick up a second job. I can handle whatever work you've got."

In that moment, shame burned hotter than the Texas sun. I felt small, like I was shrinking into the floorboards, wishing the cracked linoleum would swallow me whole.

He tapped a pen against his blotter. "From which high school did you graduate?"

My pulse thudded in my throat. "Sir…actually, I only went through eighth grade. I dropped out and enlisted. But I took the GED, and I've got the certificate right here." I showed him the crisp paper inside my folder.

He gave a contemptuous snort and tossed his handkerchief back into his pocket. "You didn't even make ninth grade? My God, boy. You think you can walk into college lectures and survive?"

I swallowed, trying to steady my voice. "I've learned a lot in the Army, discipline, tactics, solving problems on the fly, things like that. I'm not asking for an easy ride."

He tapped the desk again, scattering papers. "You don't have a rat's chance. My advice? Quit this nonsense and learn you a trade, plumbing, welding, something real." He didn't know I was already a bricklayer.

His words hit like a hammer. My chest tightened, and I fought the urge to shout back. Rage swelled, but I kept it pressed down, knowing an outburst would only prove him right.

I stifled the urge to argue. My hands itched to slam on the desk, but I clenched my fists at my sides. "So, you're saying I can't enroll? I really need to give this a try. I need a chance."

He sighed and stared at the jumble of paperwork. Reaching into his pocket again, he withdrew that same handkerchief and blew his nose for a third time. "The rules are the rules. I hate 'em, but I gotta follow 'em. At

Southwest Tech, we waive normal entrance qualifications for veterans with a GED. Stupid policy if you ask me, but you meet it. Congratulations, you can now register. But you don't belong here. You'll never survive your first semester, no matter your major, even if you pick pottery."

The contradiction spun in my head, acceptance and rejection in the same breath. My heart leapt and sank at once, the air was heavy as though the walls themselves had closed in around me.

I stood there, chest tight, the overhead lights flickering as I processed his words. The door behind me beckoned, my one way forward, or maybe the only exit.

Did I hear him correctly? Did he say he had to let me in? But then, in my nervousness, I blurted out the most foolish question, "Could you explain what majoring is?"

Mr. Staples, sitting behind a polished oak desk, looked at me with a mix of disbelief and amusement. "Oh, dear God, help me. Are you telling me you don't know what field you want to major in? A major is the discipline in which you specialize. For example, if you are an English major, your primary focus is on the English language. If you major in business, you're diving into the world of business. It's essentially what you're interested in."

I paused, my mind racing, and said something to Mr. Staples that amounted to: I didn't care what my major was, as long as the degree, when I finally earned it, erased the shame of never having finished high school. That's not exactly what I said, and I certainly didn't use the word "stigma," which I wouldn't have known at that point in my life, but that was the feeling I was trying to put into words.

Mr. Staples leaned back in his plush leather chair, the creaking sound filling the room, and shook his head slowly. I felt compelled to keep the conversation going, so I added, "I think I'm most interested in science." I didn't mention that my brother was already immersed in the physics department.

A smile, or perhaps a smirk, tugged at the corners of Mr. Staples's lips. "Science, huh?" he said, almost chuckling. "You want something that will get you booted out of here quickly, huh? Let me suggest a field of science for you. You should major in physics. That will be a true test for you. If you make it through that, and believe me, you won't, no one will ever question whether you went to high school. That's your perfect major. It will see your butt off this campus by Christmas."

CHAPTER 11

Old Main

I left Mr. Staples's office with a spring in my step, feeling as though I was floating on air. Clutched in my hand was the stack of paperwork he'd given me, my ticket to enroll for the fall 1969 semester. My heart raced with excitement. As I made my way to the red-roofed Old Main, a building steeped in history, I felt an irresistible urge to reach out and touch it. The imposing Victorian Gothic structure, built in 1903 and perched on a hill overlooking the quad, was the very cornerstone of the campus. Its intricate stonework and towering spires spoke of a bygone era when Texas State University was known as "Southwest Texas State Normal College." Back then, Old Main housed an auditorium and chapel on its second floor, complete with a grand cathedral ceiling, stage, and balcony, echoing the voices of generations past.

In the early days, when Old Main was merely an ambitious dream on blueprints and sketches, constructing such a landmark required more than just vision; it demanded a solid foundation. Workers faced an unexpected obstacle right from the start. As they labored to transform architectural plans into reality and poured concrete for the basement footings, the mixture mysteriously vanished into an abyss beneath the carefully measured forms. It was as if the earth itself had swallowed the liquid cement, refusing to hold it. Frustration mounted until they realized a hidden cave lay beneath the site. This cavernous void beneath the hilltop plot demanded careful attention, insisting that it be filled and stabilized before any truly majestic structure could rise above. Once the cave was finally addressed with loads of broken stone and packed clay, Old Main took shape, a testament to the perseverance and determination of its builders. Soon, its red roof and stone walls stood proudly, visible for miles, overlooking the growing city below. It was a symbol of hope and permanence, ready to educate generations of students, including me.

I wasn't satisfied with just brushing my fingers along Old Main's exterior; I had to step inside. My footsteps reverberated through the corridors; each echo was a reminder of the history that filled these halls. I lingered, taking in the ornate woodwork and the flow of students around me. They were diverse, with different backgrounds and stories, yet I felt an undeniable connection. I was part of this living, breathing tapestry of history and culture, eagerly anticipating my first classroom experience since the tumultuous days of junior high.

Within a day or two, I stood registered, holding a list of courses for my inaugural semester as a budding physics student. My schedule combined remedial subjects with essential college courses, a blend that both excited and intimidated me. I would be diving into English, mastering the art of communication in Fundamentals of Speech, exploring the past in History, and challenging myself with two math courses: College Algebra and Plane Trigonometry. The hurdles loomed large. I was neither a strong writer nor speaker, and the realms of algebra and trigonometry were foreign territories. But I was ready to embark on this academic journey.

I staggered through that first semester by outworking everyone else on campus. My room at Deco's old house on North Street had a single, flickering bulb over a metal desk littered with coffee cups and crumpled scratch paper. I'd snatch catnaps on the stiff twin-sized mattress, wake up to the buzzing hall light, and dive back into trigonometric proofs until my eyes burned. I'd lived on the Davis Ranch near Leesville, Texas, as a child in a rickety, paint-chipped shack where the wind whistled through broken windowpanes. Riding the rattling school bus, I shared seat 12 with a skinny kid named Wilbon Davis, both of us squashed against backpacks and stray dog hair.

Imagine my shock during that very first semester at SWTSU: Wilbon Davis, whom I once shared a lumpy school bus seat with on the way to Leesville, sat cross-legged at the front of my 8:00 a.m. trigonometry class, with the attendance pad on his lap and a devious grin bursting through his scraggly beard. He was far from the quiet, scrawny kid I remember. Wilbon, my childhood shadow from the Davis Ranch, was now tall and lean, with an unruly mass of hair cascading down to his waist in a loose braid. He wore hawkish spectacles, a pair of faded bell-bottom corduroys, and leather sandals that completed his transformation into a hippy guru of mathematics. He looked as if he'd just returned from Woodstock.

As he paced the front of the lecture hall, green chalk squeaked across the blackboard, marking his passionate and eccentric flow of sine and cosine. Wilbon's teaching style was as unconventional as his appearance. He wrote feverishly, all the while delivering quick, esoteric references to music and art alongside theorems and identities. It felt like we were diving into an unpredictable maze, where the only way out was to embrace the chaos. I struggled to keep up, furiously scribbling notes as I tried to decipher his looping handwriting. Many of the other students seemed bewildered; some even dropped the class by the second week. But I was determined to stick it out. I knew that underneath his counterculture persona, Wilbon had a clever mind, and I wouldn't let him get the better of me. I was used to struggling. What I lacked in skill, I made up for in brute persistence, sometimes staying up all night to grasp a single concept. Wilbon noticed my effort and sometimes even challenged me directly in class, shooting me a sly smile when I finally solved a problem.

Everything I'd learned about angle identities came down to the exam. I remember one test that involved a two-sided equation of twisted trig expressions that required me to summon every memory of $\sin^2\theta$, cosine double-angle formulas, and tangent sums I'd ever drilled into my skull. My pencil squealed across the page. Sweat beaded my temples and dripped onto the test sheet, smearing my half-legible scrawls.

Wilbon drifted down the center aisle, his sandals whispering against the linoleum. He'd pause behind me, peer over his glasses, and shake his head, sometimes twitching his long fingers as if he were conducting a silent symphony of disappointment. Five minutes before the bell, he snatched my exam, torn halfway through by his quick jerk, and ripped it in two. My heart stopped. Failure loomed. I tasted iron on my tongue.

Then he dropped the two halves back onto my desk, only he'd flipped them, the bottom half on top. In that instant, the fog cleared. The scattered pieces aligned like puzzle fragments. I grabbed a fresh sheet, wrote one clean line after another, and proved the identity flawlessly. The buzz of his whisper, "That's it," sent relief flooding through me. I knew it was an unfair advantage to anyone else who might need a hint, but it felt like salvation. Thank you, Wilbon Davis.

My second humiliation came weeks later in speech class, a narrow room with a single wooden podium under a harsh spotlight. For the midterm, each of us had to step up, choose a passage from a book we loved, and read aloud while a dozen peers watched, pencils poised. My palms were sweaty as I tugged at my worn copy of Desmond Morris's The Naked Ape, its spine cracked, pages thumb-marked. My voice caught on the first sentence about human evolution, my throat tightening. The classroom smelled of old paper and anxious bodies. Every eye felt like a spotlight focused on my trembling mouth. I stumbled over a word, flushed, then forced myself to inhale and continue, imagining the wind at my back.

Zoologist Desmond Morris invited us to regard humans as nothing more than another ape in his 1967 classic, The Naked Ape. He peels back our civilized polish, how we court, mate, slumber, gossip, groom, spar, and dream, placing us squarely alongside chimpanzees and gorillas. We are "risen apes," Morris insisted, clever, tireless, inventive, yet still creatures who might forget their hairy roots.

Stupidly, I had decided to read one ill-chosen paragraph. Morris laid out a chart in that paragraph comparing the penis sizes of various primates (chimpanzees, gorillas, orangutans, and humans). To my surprise, the largest penis size among the great apes is homo sapiens. I intended to surprise the class with this fact.

Thirty students filled the room beneath fluorescent lights. I stood at the podium, paperback trembling in my rough, blue-collar grip. My voice started slowing, each word echoing off the black chalkboard. My pulse thumped in my temples. As I approached Morris's anatomical breakdown, I began to quicken, breath hitching with anticipation. Then I stumbled on the fateful word: "penis." My throat seized. My tongue felt swollen. I could not force out those two syllables.

A ripple of snickering washed over the desks. My cheeks burned. The professor, clipboard in hand, cleared her throat: "Please start again, Mr. Benson?" Her tone was crisp, unyielding. Start again? My stomach lurched. I imagined myself on a construction site instead, hoisting a solid concrete block, feeling the grain of brick dust under my fingernails, and I stumbled through the sentence.

I retreated, shoulders hunched, sliding back to my seat in the last row. My damp palms clutched the desktop edge as the murmurs continued. The linoleum floor's scratched pattern blurred beneath my gaze. Every eye in the room seemed fixed on me. I curled into myself, willing the laughter to dissolve. And adding to my humiliation, my younger brother John sat three seats away, red pen poised over his notes. I dared not meet his eyes, afraid he'd be smirking, too.

By semester's end, my transcript was a collage of regrets: English: C, College Algebra: B, Plane

Trigonometry: C, U.S. History: D, Fundamentals of Speech: B. A rocky start. Considering where I started, I felt fortunate to have completed five college courses, even though my grades were barely passing. Storm clouds were threatening. My GI bill stipend, only $400 per month, was running out. My first marriage was on the cutting-room floor, and I faced court-ordered child support payments. The future looked bleak. My dream of crawling away from my redneck existence was fading fast. It looked like Mr. Staples was right. I didn't have what it takes to endure higher education.

CHAPTER 12

Rotting Tuna

Restlessness has followed me since birth, a sleep disorder shaping most of my days. Late into the night, my thoughts ramble wildly. I sometimes wonder whether other people's minds work the way mine does, jumping from one thing to the next, never stopping long enough to settle. That night in 1970 at my grandparents' house in the sandhills of South Texas is a good example. What I am about to tell you is not in perfect temporal order, but to know my life, you need to know my childhood.

I have always had trouble reading and writing. I have great spelling difficulty. The words you are reading now are made possible by tools; without them, my letters would slip. My focus wanders. Even simple choices get interrupted by side thoughts. I recently struggled with the word "thwarted" and chased it in circles, convinced it started with an "f." After hunting for it too long, I gave up and used "frustrated" instead. That is how my mind has worked since I was a boy. I can usually hear spelling and grammar mistakes, but I cannot see them with my eyes. I have an app on my desktop that helps me correct grammar mistakes, and another app that reads back what I write aloud. These crutches help, but they are not perfect.

My trouble with reading and spelling, along with my inability to sit still, caused me to repeat the third and fifth grades. In the 1940s and 1950s, no one at school knew the term ADHD. Teachers just said an afflicted student was lazy, dumb, or worse. If therapy was needed, it would come in the form of a wooden paddle or a leather belt. My teachers told my mother I was slow, maybe even retarded, and they predicted the worst for me. She believed them.

Finally, my mother pulled me from public school and put me in a Catholic school, certain the nuns would fix me with strict discipline. Harsh discipline is the right word for it. There was no allowance for fidgeting, no patience for a boy who could not hold still. If you wiggled in your seat, a ruler or stick would find your knuckles. If you disobeyed, the punishment came fast.

Each morning began with Mass. The bell sent us into the cool shadow of the chapel, pews creaking as we sat. The priest walked to the altar in a violet chasuble, altar boys in crisp cassocks swinging a thurible that trailed a ribbon of gray-blue smoke. That incense clung to your hair and clothes. When the priest said, "Dominus vobiscum," the Latin rolled over us like a tide. I leaned forward, wanting to understand every word. Even now, I regret never learning to read that ancient dead language.

Our classrooms were inside Mission San José, its honey-colored limestone looking out toward the San Antonio River. Built in 1720, its buttresses and bell tower carried the past on our faces. I pressed my hands to the cool stone and imagined the Papopa, Pastia, and Soluajan tribes kneeling there as priests raised crucifixes. Within the same walls, I learned long division and history from nuns whose black-and-white habits rustled in the drafty corridors. I never converted; I was more a student of buildings and stories than of doctrine.

Fridays at lunch were fish days. Sister Agatha's voice told us to clean every crumb from our metal trays, but one day, I remember, the fish had gone bad. When the line of nuns turned away, trays of tuna sandwiches became a problem to solve. Some kids dumped their fish into the barrels. Others pretended to nibble and flicked vile mucus under the tables.

Juan Ríos leaned into me and whispered, "Bobby, you taste that?" I bit and recoiled. The rancid stink of canned tuna hit me hard. I washed my mouth with lukewarm water and gagged. All around, kids slid pieces of sandwich into the trash or hid them under the table. It was one of the hardest things to do, but I swallowed the last bite of my putrid sandwich and pushed my chair back to deposit my tray into the wash hole. My chair legs screeched across the scuffed floor. A half-eaten tuna sandwich flipped through the air and landed directly on my tray with a wet thud. My heart pounded. I turned, searching for the culprit who had thrown it. No one owned it.

Footsteps came up fast behind me. I turned and saw Sister Frances, horn-rimmed spectacles, fingers around her yardstick. Her habit brushed the floor. "Master Benson," she said, "stop there and sit."

My stomach knotted. "But Sister Frances, I..."

"You will sit," she said. She tapped the tray's raised edge where the sandwich lay. "And finish that sandwich."

I tried to explain. "Sister, I already ate mine, honestly," I said. "This one has bites on it. It's covered in slobber. I could get sick."

Her lips pinched. The stick hovered. I sat, rigid, staring at two soggy slices pressed around gray tuna and streaks of mayonnaise. My palms sweated on the cold table. The lights hummed. My throat tightened. The metallic taste crept up.

Her stick cracked my left hand. Pain jumped up my arm. "Start eating," she said. "Or you'll feel this on your other hand, too."

I lifted the sandwich. Each bite tasted like brine and rubbery fish. I chewed, wanting my body to behave. My stomach coiled. I forced it down until half was gone. "Is that enough?" I said, my voice high.

She shook her head. "Take another bite." The stick tapped the tray.

I bit again. The room spun. My pulse thundered. I fought the heaves and lost. With a wrenching sob I threw up onto the table and floor. Green-yellow liquid pooled, lumps of soggy bread bobbing. I doubled over, shaking.

Sister Frances grabbed my arm and pulled me up. She scraped the table with the yardstick and set my tray back in place. "Don't you dare step in that mess," she said. "Go to the bathroom. Clean yourself up. Then back to class. And stay until lunch is over."

My face burned. My hands shook. The room came back in pieces. Kids looked away. The smell of tuna and vomit clung to everything.

That afternoon, my mother picked me up; Catholic school had no bus to take me home. I told her I had slipped in another kid's vomit. I did not say anything more. My mind replayed the slap of the stick, the sour tang of fish, the puddle at my knees. From that day, I could not bear to eat after anyone. I flinched at shared desserts. At Thanksgiving, I watched relatives dip spoons into the dressing and knew I would not touch it again.

The years that followed did not ease my school troubles. Reading stayed slow. Spelling slipped. Concentration was a worn rope that snapped every time I tugged at it. The nuns believed in pounding knowledge into boys, and I carried the marks on my hands and deeper, where no one could see. There were small mercies. A teacher once let me draw when my letters would not cooperate. A classmate shared his marbles without mocking me. Little sparks in a dark year.

I think of my mother in those days. She worked long hours, hoping that discipline would help me learn. She was trying to keep me from slipping farther behind. If her choices sometimes hurt, it was because she wanted me to rise. I did not make it easy on her.

I still hear the ruler's crack and the priest's Latin rising and falling. I still smell incense and bleach. The cafeteria is bright in my mind, with metal trays, the echo off tile, kids whispering when the nuns turned. The tuna sandwich sits there, and my throat closes even now.

That day did not start my fear of other people's food, but it deepened it. From then on, anything that looked handled set off alarms. It was not a quirk I could outgrow. It was a rule my body enforced.

Looking back, I see how fear attached itself to ordinary things. For most kids, lunch was just lunch. For me, it became a test I failed in front of everyone. That is how some memories work. They fix themselves inside you and become the lens through which you see every version of the exact moment.

The rest of the day after the tuna incident passed in a blur. The yardstick, the bathroom, the shuffle back to class, the ride home. I did not sleep that night. When I closed my eyes, I saw the sandwich on the tray and felt the room tilt. By morning, I knew this was not going away. It had become part of me.

Years later, whenever I entered a cafeteria, I would scan for exits before looking for a seat. I kept a hand on my tray as if someone might tip their food into it. At family gatherings, I drew my plate closer, not to be rude but to keep in control. People laughed from time to time; I did not mind. They had not met the yardstick.

There were good days mixed in, even then. I could ride my bike and feel the wind in my face. I could find a quiet spot outside and listen to birds. That steadied me the way nothing at school could. But the tuna scene stayed, the way certain things stay when they find the soft spot and press.

It is hard to explain to anyone who has not felt it how your body can refuse what your mind begs it to accept. I wanted to be the boy who shrugged and ate, who brushed off the smell, who swallowed and moved on. Instead, I was the boy who got sick and kept a rule for the rest of his life.

That was the week I realized that some battles aren't fought in the head; they're fought in the gut, nerves, and skin. Once you lose them, you can't go back to who you were. You learn to live with the rules your body sets for you, and you build your days around it.

I am not proud of those school years, but I am not ashamed either. They are part of the path that led to this point. If I had been a different kind of boy, this story would be different. I was who I was, in the time I was given. The cafeteria, the chapel, and the classroom made their marks, and I carry them still.

CHAPTER 13

Science, Huh?

As a child, I'd discovered some old books about radio technology in our house. The drawings in these books were fascinating. I tried unsuccessfully to understand the writing and diagrams. These were old texts that belonged to my father. He had attempted to learn electronics after his wartime battles, but I don't think he knew much. Thinking about birds and radios offered an escape from Dad's unpredictable, sometimes violent behavior.

The concept of transmitting human voices through empty space and across continents filled me with awe. Long before I knew about Ampère, Gauss, Faraday, Maxwell, Hertz, or Marconi, I was mesmerized by radios. I was born at a time when radio was king. Television would have been unimaginable for normal people. A child like me could peer into the backside of a radio and see the glass tubes with their dim orange filaments glowing. You could see condensers, coils, and tuning devices that looked like rows of tiny metal discs with cutouts. How could something like this work? It was spiritual to me.

Back then, home radios were furniture. My mother's radio took up the space of a chair; a great blonde box with a curved top, majestically sitting in the living room. I raced home from school every day to listen to my favorite kids' program, "Bobby Benson and his B-Bar-B riders." I guess you will know why. I'd listen to everything that was broadcast just to be close to the magic. I was glued to my mom's radio when Dwight D. Eisenhower was elected president in 1952. But the most thrilling part of all was the radio's shortwave band. This is where I could slowly turn the dial and hear stations from all around the world. Aside from some humans in my life, radios may have been the most influential element I encountered as a child.

I don't remember my exact age, but I was under 10 when I first met another kid who had his own radio. But it wasn't like the big one in our living room. He had a crystal set. It was all his, and he had it in his bedroom, hooked to a long copper wire that stretched into his backyard. All the parts for this little radio were mounted on a piece of wood. There were no tubes or speakers, and you didn't have to plug it in. I could see all the simple parts. He tuned in to different stations by moving a strip of metal across a bunch of wire wrapped around something that looked like a short piece of broom handle.

He tuned it to his favorite station and let me listen. When I slipped the headphones over my ears, I could not believe what I was hearing! Mom's radio at home was guarded. I was not allowed to turn it on without her permission. The thrill and empowerment of this little device almost took my breath away. I had to have one.

As soon as I was home, I began begging my mom to buy me the parts so I could build my own crystal set. My friend had given me a long roll of copper wire for the antenna and coil. I could find some wood and cut a piece of metal from a spinach can for the tuner. The only things I needed were a pair of headphones and a crystal of Germanium. Oh, I would also need a cat whisker. A cat whisker was a stiff length of wire made from flexible steel. It was used to probe the crystal's surface and find a spot where a radio signal could be detected. Most of these parts were cheap and available at the local Hobby Shop. But headphones were a different story. They would cost dollars, not pocket change.

"Mom. I want to build a little radio like my friend at school has. It's a crystal set, and the parts wouldn't cost too much. Tommy told me I could get some headphones at a store on South Presa, called an Army Store."

She told me maybe, if it didn't cost too much. I was thrilled. That Saturday, we stopped at the Surplus store and checked on the cost of headphones. The cheapest pair was priced at $5. This was devastating news. She immediately said we could not afford headphones. Five dollars then was like $65 today.

Mom looked at my sinking face with sad eyes. I wasn't crying, but almost. It may have been the first time in my life that I realized that we were poor. It hurt. It seemed very unfair. Mom had a big radio with glowing tubes, but I couldn't have a crystal set. It occurred to me later in life that Dad probably acquired her blonde radio illegally. I never knew for sure, but I no longer felt the hurt. Instead, I felt Mom's hurt. I know for sure that she was deeply wounded, that she could not give me the object that I so passionately craved at my young age.

It is my nature to quickly get over negative events in my life. I never forgot about crystal sets, but I didn't dwell on them for more than a few days, because we couldn't afford the parts. Life went on.

CHAPTER 14

Uncle Louie

Humans are so very different. The adage about "nature or nurture" shaping a person's psyche has never been definitively resolved. Is personality a genetically determined seed? I think so. The environment in which that seed grows is the soil, water, and sunlight. The same seed planted in different conditions can grow into widely different expressions of the same core design.

I was born magnetically shackled to the Earth and all its extraordinary wildness. My "raising" only magnified this proclivity and deepened my sense of isolation. I form friendships with great difficulty and tend to shy away from social interactions. Only my close family knows this about me, because I've learned to fake social skills by mimicking others. I know what to say and how to say it, and I charm unsuspecting acquaintances, but it's all complete social camouflage. Not everyone is fooled, but most are.

I share some traits with sociopaths. I looked it up. A sociopath disregards laws and social norms. I did that in my youth. A sociopath lies and is deceitful (chronically). This fits, but my behavior here is not chronic. A sociopath is impulsive and aggressive. That's me, but I'm not physically aggressive. A sociopath lacks remorse or guilt. I strongly feel remorse and regret. A sociopath has difficulty forming genuine emotional connections. Okay. That's me, but I do have emotional connections with my close family and my few "friends." A sociopath is often manipulative and charming, possessing a certain charisma. That probably fits, too. This is a biased self-diagnosis. I have no idea how others would see me.

What is not on this list is the capacity to love other humans. I have maintained a lifelong struggle with understanding what love truly means and whether I am capable of loving. I feel the need to be loved. That need is powerful. I'm very empathetic, and it shatters me to know people are hurt, sometimes by me. But is empathy love? There are people in my life who would be crushing losses if they were gone. But is that love, or selfishness? Love is too complex for me to understand, and I've tolerated a lifetime of discomfort because this issue is not settled for me. Is this question settled for others?

Sometimes when I came to visit my grandparents, I'd pass through the town of Luling and smell the odor of sulfur. The sulfur smell of Luling would bring another memory of a fitful night at my grandparents' home in the winter of 1970, between the fall and spring semester of college in San Marcos. The memory was of my grandfather, Palmero, or Grandpa, as we called him.

South of Luling, the next small community is Belmont. The landscape begins to change as you drop off a steep hill onto the floodplain of the Guadalupe River. You'll recall the triangle I described earlier, the fishing camp on the Guadalupe, Capóte Knob, and the old ranch house near Leesville, that enormous physiographic shape that was also, for me, a haunting emotional one. The first of those points, the fishing camp, is where this story belongs.

The first point on my triangle was a fishing camp just east of where Highway 80 crosses the Guadalupe River. You can see the camp as you cross the bridge. Now known as Garcia River Campgrounds, the site has undergone significant development since I spent nights there fishing and sleeping. My grandfather used to take my brother and me there to teach us how to fish and camp out, sleeping wrapped in blankets on the ground by a warm and flickering campfire.

Grandpa had only two fishing spots that I knew of: this campsite on the river and a private property on Willow Creek near Leesville. I have vivid memories of my adventures with my homemade cane pole and wriggling worms on my hook. I can't assign a date or even a year to these adventures, but both my brother, John, and I were kids, perhaps too young to be in school. Grandpa was a master storyteller and beguiled us kids with tall tales of ghost encounters and family lore about our ancestors from Cuba. He had a condition that made him snort every few seconds as he laid out his narratives.

After wolfing down slabs of river-caught catfish, Grandpa would snort a few times and start up a story. One of his favorites was his first childhood encounter with a ghost, which he swore had happened just about right here, where we boys were wrapped in our blankets by the fire. He cleared his throat, snorted a couple of times, and began.

"Well, boys, it was many years ago when I was a young man. Back then, my brothers and me lived in Caldwell County. That was in the late 1800s, before I met your grandma. We came right here to catch us some channel cats. Snort. I'd settled off to sleep wrapped in a wool blanket, snort, just like you two are now. The fire had died down, and I was warm and cozy-like and fast asleep."

He tossed another chunk of wood on the fire, and "snort", continued. "We were all settling lard-fried mud cats in our, snort, bellies and was dead to the world, asleep. Then's when it happened. I was woke up by the sound of something walking in the dried-up pecan leaves, snort. I was laying on my belly and started to get real scared, 'cause the steps was getting closer. I stayed real still and kept my head under the blanket. The steps stopped coming, and it was really quiet for a long time. I didn't move a muscle, snort."

Grandpa cleared his throat, snorted, and did not continue the story. He stayed quiet and watched the fire. Johnny didn't say anything either, and after a while, I couldn't stand the silence and spoke with a shaky voice.

"But Grandpa, what was it making the steps? Did you see something?"

He cleared his throat again, "I thought you boys would want to know what happened next. Snort, like I said, I was on my belly and had my face covered with the blanket. Then's when I felt something. There was a hand reaching under the blanket, and that hand grabbed my foot. I wanted to jerk away, but I couldn't move, snort, I was frozen up. The hand raised up my foot, held it there for a few seconds, and then dropped it. I heard it make a low-sounding grunt, but it weren't like a human voice. It was like a hollow sound you might hear in a cold, dark cave. Then it walked back away, rustling in the dry leaves again. Snort, I've seen many a ghost in my old life, but that was the only time one came up and raised my foot. I don't know what it wanted or why it did it, but I was mostly frozen-scared. Snort."

Just being a little kid, I was scared to hear this story and firmly believed everything Grandpa said. I never asked Johnny, but I'm sure my brother was afraid, too. To this day, when I have to cross the river at Belmont, that story floods my mind and makes me smile.

As I grew up, the Benson family had a profoundly negative influence on me. I didn't know much about my paternal genealogy, except for the relatives who were alive and with whom I was occasionally forced to interact. As I have expressed, I feared my father the most, because his wrath disrupted my life the most. My youthful reaction to his fractured personality was a deep-seated need to please him in some way. I never succeeded.

My mother's family, the Palmeros, were completely unlike the Bensons. They were respected, stable, hardworking, and safe people. I never once felt afraid around the Palmeros. I loved to say my Grandpa's whole name. He was "Louis Mariano Palmero," but most everyone, all the neighbors and family, called him "Uncle Louie." He grew up during a time and in a place where people of Hispanic origin, and with Spanish-sounding surnames, were discriminated against and thought of as the lower classes. Grandpa told us, and everyone else, that his family originated in Italy. He did not want to be mistaken for a Mexican.

I believe this throughout my childhood and young adult life, that our Palmero family, who immigrated from Cuba in the late 19th century, was of Italian origin. I was a little doubtful about this, owing to his and my mother's light green eye color, very light complexion, and my mother's blonde hair. Later in life, when I had the opportunity to be in Rome, I curiously paged through an extraordinarily thick Roman telephone book, and my memory is that I discovered no one with the surname of Palmero. It seemed vanishingly unlikely that the Palmero line originated on the Italian peninsula. My Ancestry.com DNA test and a 2025 AI deep search provided new information about my Palmero relatives, previously unknown to my mother's family, suggesting that my Palmero family originated in Spain and then immigrated to Cuba in the mid-1600s. They were a prominent family and were in the Sugar Cane business.

After the Cuban 10-year war, in which their father was executed by firing squad, my great-grandfather, Jose Rafael Palmero, immigrated to New York in the 1860s and became a civil engineer. During his professional life, José built bridges along the Mississippi River, got married in Louisiana, and eventually settled in Caldwell County, Texas. His son, my grandfather Louis, exerted an enormous influence on my life, both genetically and developmentally. Thank you, Grandpa.

CHAPTER 15

Hercules

I pulled into Leesville, which could have been the smallest place in Texas that people called a town. Here, there was an intersection on Highway 80. You could turn left to the east and go toward the old house where the great turkey annihilation occurred. I tried to block that vision from my mind.

On the corner was the only business in Leesville, Mr. Kidd's little store. Mr. Kidd was a scary man. He was a tall, skinny man with a loud and harsh voice. Some undisclosed accident or other misfortune had ripped out his left eye, and he didn't have the courtesy to wear a patch over the sunken hole in his face. We feared Mr. Kidd. Afraid, but not in the way Dad scared us with his bouts of violence. I don't think my father hated children, but Mr. Kidd certainly did. Maybe the loss of his eye was connected to some event that had to do with children, or why would he hate young boys the way he did?

Looking back, I see how fear came in different shades. One fear demanded obedience, the other left you hollow. Both etched themselves into the way I measured people for the rest of my life.

Now that I had grown to be a young man and was pondering a possible significant change in my life, the thought of Mr. Kidd did not scare me. I think he was probably dead anyway, and nothing was left of his store but a falling-down shack at the intersection.

Here, you could also turn right and follow a deep sandy road to Mimmy and Grandpa's place near Dewville, only about seven miles. I loved being alone along this sandy road, not only because I loved being with my grandparents, but because it was such a wild place. There were no houses I can remember, and encountering other drivers was rare. I often stopped where the road crossed Sandies Creek, usually dry, and listened to the birds calling and singing. There were never many. But today I was running late, so I didn't stop at the dry creek crossing. In another ten minutes, I pulled up to their modest frame house on Lakey Road. They weren't expecting me, but I knew I was always welcome.

As soon as I stepped out of my old car, a familiar scent was in the air. Grandpa, ever the blacksmith, was at his forge, no doubt preparing to hammer out a new part for his little homemade lawn tractor. If you have not smelled the smoke from a searing red-hot blacksmith's forge, it has a sharpness to it, almost bitter, like scorched stone and iron dust mingling in the breath of the flame. It smells and looks ancient. It transports you to another time. But in this narrative, I won't transport you all that far back, only to a time that made me sure I had to escape my crazed father. Another time and place where I remember the smell of a blacksmith's forge, at Dad's riding stable, where farriers would often visit to shoe horses. They needed to heat the metal horseshoes on a forge to make them flexible enough to hammer them to fit the hoofs.

Even now, when I catch the faintest smell of burning metal, I am transported back to those years. Memory rides the senses, and scents have always been the strongest guides into the past.

I think I was 16 years old, about to turn 17, immature, and skinny like a fence post. I weighed around 120 pounds and was not very strong. Our family was living in Pasadena, in the southeast part of Harris County, a suburb of Houston. Somehow, my father had acquired some substantial chunk of money, most likely by some criminal act. He planned to invest that money in a business that was surely going to make him rich, his lifelong ambition. Remember, he loved horses.

The 1400-acre Memorial Park, a former military property during World War I, lay at the western edge of downtown Houston. Adjacent to the park was a riding stable where the wealthy residents of River Oaks boarded their horses. Riding trails spiderwebbed the pine trees in the park, and on the weekends, the trails were well used.

Miraculously, our father acquired that riding stable. He either leased it or bought it. I don't know which.

My designated tasks were to rake horse manure out of the stalls, brush down the horses, and saddle them when Dad got a call that one or more of the horse owners decided to drop by for a ride. These wealthy, well-dressed, well-spoken, upper-class, super-rich kids were astonishing to me, a class of people I had never encountered. Some of the rich kids, mostly boys, were wild and indulged in all manner of unacceptable behaviors that their parents either tolerated or didn't know about. As you will imagine, even though I was about their age, I was not included in their social group.

Aside from Dad's income from boarding horses, he also rented horses by the hour to other lower-class people who dropped by and wanted to ride the trails. If the folks were inexperienced riders, I would be assigned to lead the group and teach them how to handle the old, worn-out, and saddle-sore nags that served as rent horses.

Dad tried to settle into running the marginal, low-profit venture, but a few months of this was all he could take. He needed more excitement. He decided to build a rodeo arena and get into the big money by putting on a show every weekend, saving the bull riding, the most anticipated, for the last act. There would be plenty of beer at the concession stand, because he wanted the fans to be tipsy and not leave early. I don't know how he accomplished it, but within a few months, the pens, the arena, the lights, and the bleachers were completed, and

Dad rented a string of bucking stock, wrestling steer, and roping calves. Of course, he wanted the most vicious, dangerous snot-slinging bulls he could find.

It was time to put on the inaugural show. I remember that my school had already let out for the Christmas holidays. It was going to be a two-night event, Friday evening and Saturday evening, and he wanted it to thrill the audience. He printed up a big pile of posters advertising the event and posted them in the windows of every building and business that would allow it.

Three large cattle trucks delivered the riding stock on Friday morning. I must admit, I had become excited about this by now and was anticipating the show with great enthusiasm. Enthusiasm until Dad called me over with a shit-eating grin on his face.

"Hey there, Bob-O-Ree-Bob. I gotta little chore for you. I picked out the smallest one, so I don't think you got much trouble coming."

At first, I didn't have a clue what he meant, but I did have history. When he grinned that way and used his fake upbeat personality, it certainly would be some job I would not want to do. Maybe he wanted me to grease the hinges on the chute gates. He could want me to turn the water on at the trough so the bucking horses could drink. Maybe it was something every boy would like to do, like getting on the tractor and dragging the disk harrow around the arena so the cowboys would have a soft place to land when they were ejected from behind the hump on the back of a wild-eyed, slobbering, snot-dripping bucking bull. But what did he mean about "the smallest one?" I fidgeted and put my head down.

"Now, don't you go scaredy cat. This is going to be the best fun you ever had. Now get that head up and act like a man. Goddamnit, Bobby, you're a Benson."

I did raise my head, but by now, I realized that whatever he wanted me to do, I wasn't going to have a choice. He grabbed my arm and more or less dragged me toward the chutes. I tried to walk fast enough, but with him pulling my arm, I struggled to keep my balance.

Then, a bolt of lightning struck. As we rounded the corner of the horse barn, I saw my fate. Lying down in chute number three was an enraged bull, bawling and swinging his horns from side to side, frantically searching for a way out. One of the men, who had trucked the bucking stock in, was stabbing the prongs of a hotshot right into the flank of the gray, humped-back monster. Every time the poor animal was electrocuted with the hotshot, he would spit and bellow and try to gain his feet, but with one horn between the chute-gate boards, he couldn't manage it.

Now Dad became enraged, "Son-of a Mother-Fucking Bitch," he yelled at the man trying to get the bull up, "I'm paying your ass to handle your goddamn stock. I gotta see how good these bastards are gonna buck."

Dad released my hand. I stood frozen in terror. He hadn't directly told me, but I knew my fate. He was going to make me get on that bull's back and try to ride it. I had been around horses, and cows, and goats, and chickens, and pigs, and all that stuff. I had been to rodeos and seen grown men dragged around an arena because their hand was tangled in their bull rope. I had seen them stepped on by 2000-pound bulls, and hauled away in ambulances. I felt like crying, but I held my tears back.

"Daddy, Daddy, I don't want to. I'm scared. I don't want to."

"You sure as hell are going to climb up on that chute and get on his back." The slabbering stinky animal was finally standing up. "It's time you learned to buck up and deal with things you don't want to do. Hell boy, I was hanging on to bucking horses when I was five years old. Your Grandpa made me do it, and I was scared, too, but it taught me some good lessons. What do you want? To grow up and be a sniveling girly pussy boy? Ain't no son of mine going to be prissy and acting like some faggy queer."

For a few seconds, my world seemed to move in slow motion. I surveyed the trapped bovine, on his feet, kicking the slatted two-by-six pine board that made his small jail, confining him to that small space. He was mostly gray with an exaggerated hump on his shoulders. From my point of view, it reminded me of a dolphin's fin, but much fatter. The tips of his horns were cut off, and his whole rear end was smeared with liquid feces. It looked like he'd had the runs and was using his tail to wipe his butt. His tail was dripping with a foul, brown liquid, manure embedded with pea-sized chunks of feces.

The stockman propped his hotshot against the fence. He draped one end of a rope just behind the shoulders of the beast. The rope had a brass, dull-sounding bell dangling from it, and my dad used a stiff piece of hooked wire to reach under the bull and pull the rope around its body just behind its front legs. Another smaller rope was pulled around the bull's flanks. This one was meant to make the animal kick higher once the monster and I were released into the arena.

I've seen plenty of bull riding but never thought about how terrifying it would be to do it myself. Usually, there was a stupid-looking clown in the arena, whose job it was to distract the bull if the rider was bucked off, so he would not be gored to death before he could run and climb the nearest fence. There was no clown out there to save me.

He was pushing me to climb to the top of the chute, where my fate was waiting to experience raging Zebu vengeance.

"Now, Bobby, get your skinny ass up there and put your right hand under that rope. And do it fast, so that bastard don't lay down again."

My world was spinning, and I felt sick to my stomach, but I found myself on the bull's back with my hand holding the tightened rope and waiting for the stockman to unlock and sling the chute gate open. I heard the clanking of the metal release hinge as the bull and its rider (me), leaped from confinement, twisting and kicking. I turned my head, a big mistake, and saw my father jerk hard on the flanking rope and cinched it as tight as he could.

I didn't last very long, just a second maybe, but it seemed like minutes. The bull bucked in a merry-go-round, tight spin, and my short legs and weak right arm could not hold a grip on Hercules's (that was his name) loose skin. With the second high kick, I was loosened entirely from the rope and the animal. Flipping backward in the stuffy air, several feet above the arena dirt, I could see the shitty butt and elevated legs of my Nemesis. It seemed like forever before I felt the impact of the arena floor crushing my chin into my chest. I did not go unconscious, and could see the gray, hump-backed locomotive, head down, blunted horns at eye level, coming for me. I scrambled to my feet and started running for the fence. If not for the bravery of the stockman who ran in front of the chasing bull, I might not have been here today to tell you this story. I made it to the fence and safety.

In those frantic seconds, I felt what it meant to want life more than anything else. Survival, not pride or defiance, was all I carried in my bones.

My father was bent over in laughter. I've never understood how he could find humor in other people's fear and pain. As I limped out of the arena with a twisted ankle, Dad stopped me and made me sit on the bleachers. He pulled out a little can of what he called medicine, which was liquid ether (he called it Ether Squibb). He poured it all over my swollen joint. It felt like vaporous ice, but the pain was gone, temporarily. I felt better. It was a bad day, but the next day was going to be worse.

CHAPTER 16

College Boys

My ankle was hurting badly the night after the bull ride, so I didn't get much sleep. I was in bed at our Pasadena house. I had no idea what time it was, but it was still dark. The wind blew as sporadic rain fell outside. At times, big drops would pelt the roof, and the wind would pick up in the downpour. I could hear Mom and Dad in the kitchen, and Dad was cursing about the weather.

"Goddamnit, I don't need this shit. I spent all that cash, and if this rain doesn't stop pretty soon, ain't nobody's going to come out for the show tonight, and all that dirt we hauled in is going to be nothing but a big mud pile."

I didn't hear Mom answer. She was probably cooking breakfast. My ankle was throbbing with pain, and I was wondering if Dad would pour some more ether on it. That made the pain stop for a little while, but the fumes made me feel sick. I dreaded the day ahead of me. It wouldn't have mattered if my leg was totally broken and flopping around, as long as he had a can of the cold ether to put on it, I'd be expected to work. I smelled fried eggs.

Dad opened the bedroom door and yelled at me to get up and get dressed. Johnny and I shared the bedroom, and he woke and rubbed his eyes. It wasn't fair, but I hoped that Dad would make Johnny get up and go with us. Although I was usually Dad's target, sometimes he spread his anger between us, and when he focused on Johnny, that bought me a little time to recover from attacks on me.

I got up, hobbled to the bathroom, and splashed cold water on my face. I put on my pants and a shirt. I heard Dad yelling at Johnny to get his butt up.

"Bobby and me are going to the stables early, and you and your momma are coming later. I need her to pick up some stuff at the store. As soon as Bobby's out of the bathroom, you get your butt in there and get ready."

I got my tennis shoes on, but I struggle to put them on my bad ankle. I loosened the lace as much as I could. I was barely able to tie it.

I could see a little light in the gray sky through the bathroom windows, and it looked like the rain was slowing down.

I only ate a slice of bacon and part of a pancake before Dad, and I started for the arena on the other side of Houston, a drive that usually took at least an hour. He didn't offer to pour any ether on my foot, but I wish he had. The bouncing of the beat-up old Chevrolet pickup shot daggers up my leg. It was still barely daylight, but Dad pulled a bottle of Vodka from under the seat and took a long swig. My mother said that he believed that if he drank Vodka, people couldn't smell the alcohol on his breath. He also kept a stash of Benzedrine, or Bennies as he called them, and popped those down multiple times per day. I think that during the war, the Army gave Bennies to soldiers to keep them focused and alert.

By the time we arrived at the arena, the Vodka and the Bennies were taking effect. I knew the phases well. At first, he would be jolly and make jokes. Usually, it was a joke I'd already heard. I always tried to smile and respond with the best fake laugh I could. He was so high by then that he wouldn't notice my laugh was fake. He'd look over at me with glassy eyes and grin.

The next step would be his silent treatment. He would look straight ahead, stare at the road, and never turn his head. He intended to give the impression that he was off thinking about something completely unrelated to the two of us bouncing along in the old truck, but by experience, I knew what was coming next. Sometimes I would wait for a couple of minutes. Sometimes it would happen after several silent minutes. But the rattlesnake move was inevitable.

WHAM. Even though I knew it was coming, it never failed to startle me. Like a lightning bolt from a towering Cumulonimbus cloud, his sun-roughed, brick-worn, short-fingered right hand would tunnel from its resting place beside his leg and grab my thigh with an iron grip. Even though I always knew it was coming, it happened so fast that I was still startled and reacted as he had hoped. He'd roar in laughter. I never joined in his roaring guffaw, because history had convinced me that the next phase of his progressive intoxication would devolve into something sinister.

By the time we drove into the parking lot at the arena, the weather had changed entirely. Even though the ground was muddy and slick, the sky had completely cleared, and a chilling Texas norther was howling. I needed a coat but didn't have one. I glance at the nasty bulls milling around in the holding pens, but I cannot see the one I'd been forced to mount yesterday. Someone else would have that pleasure today.

The two college boys, big, strong types, that Dad had hired as helpers, were standing beside the barn, smoking cigarettes and avoiding the chilly wind. They looked like they had been drinking all night; they were not in the mood to spend a day at hard labor. Dad waved them over so he could explain the first task he wanted them to complete.

Dad got pissed off the instant he noticed the way they were walking. "What the hell. Are you two lazy bastards drunk?"

They were slack-faced and pink-eyed. "Well, we were cold waiting for you, so we took a little nip to stay warm. But we're not drunk, no drunker than you look, yourself."

I knew that sort of backtalk and attitude was something Dad was unlikely to tolerate. "Shut your goddamned mouths. You don't know shit about what I've been doing, and it ain't one bit your business. I'm about half ready to run you off this job anyway. See that faucet over there? And those two buckets? Get your asses over there and carry water to that trough in the bull pen. I want that trough filled up. You got that?" And that ought to take about 15 minutes, so get started."

Dad said to follow him to the barn and wanted me to haul hay to the pens and toss it over the board fence for the roping calves. He was grumbling about the drunk college boys. I could see looming trouble. He didn't think they were moving fast enough, and his anger was escalating. By the time I got the first load of hay to the calves, I could hear him arguing with the boys, a heated argument, too. I could not hear exactly what was being said, but I saw one of the boys grit his teeth and sling one of the buckets across the ground toward the bull pens. Dad was red-faced, pointing toward the rolling bucket, and ordering the blond one to retrieve it and return to work. Dad was furious and screamed for them to get their stuff and get off his property. Now, they were all yelling so loudly I could hear everything.

"Okay, old man, we don't need this job anyway. You can carry your own fucking water. Just pay us for yesterday and this morning, and fuck your damn bulls."

"I'm not paying you lazy bastards a damned penny. You've got five minutes to get your shit and get on the road, or I'll show you what you call an "old man" can do with your pretty-boy faces." The dark-haired, skinny boy staggered a little and pulled a small pistol from his back pocket.

"You sure as hell are going to pay us, or I'll show you what I can do to YOUR face."

Just as fast as Dad had grabbed my leg in the truck, he slapped the gun out of blacky's hand. It fell in the mud, and Dad stepped on it, pushing it under the reddish slush.

"No matter what you do, we are staying right here until we get our money. You might think you can take us on, but there are two of us, and we'll push your face down in that mud right there with Billy's pistol. You got that old man?"

Then, what Dad did surprised me. He turned his back on the boys and walked straight toward me. I was halfway to the calf pen with my second load of hay.

"Bobby, stop right there. I got a new job for you."

He stopped and plucked a flat pint bottle of Vodka from his back pocket and took a long draught of the clear liquid. "Those two bastards need an ass kicking and they need it bad. I could do it easily, but since they're still considered kids by the law, they'd put me in jail for it. So, you need to go and do it, instead. The law won't do nothing to you, since you're a kid, too. So go find a stick or a two-by-four and run them off the place. They can't do much against you, 'cause they're stinking drunk and can barely walk upright."

These were shocking words to hear. These boys were grown-man-sized, and I weighed about 115 pounds, and I was limping on one leg. I started shivering at the thought of what Dad was telling me to do.

Red-faced, Dad yelled at me. "Now do what the hell I told you to do. We ain't got time to let this go on. Do what I say, or you'll have a whipping coming, too."

"But, Daddy, I'm scared. Please don't make me fight those boys. Even if they are drunk, I'm still afraid." I could feel a knot of sickness in my stomach, and my shoulders were shaking.

"Okay, you little bastard. I always knew you were a bastard, and not a kid of mine. You've got no guts like a Benson has guts. Or maybe you're a queer. That's what makes you afraid to fight, like a girly boy is afraid to fight. There ain't no fear in a Benson, so you ain't one, that's for sure. You're just a snotty-nosed coward. You get the hell out of my sight until I call you, you hear me?"

Shaking and with tears in my eyes, I hobbled over behind the barn and climbed into Dad's truck. I was sniffling and crying sometimes and thinking about the whipping I was sure to get as soon as he got those college boys off the property. I hated myself. I was too old to cry. And maybe Dad was right. Maybe I was a coward, and perhaps I was not a Benson. Through my sobs, I could hear yelling and screaming from the arena.

Then, a stupid and dangerous idea came to my mind. Maybe I couldn't fight, or I might not be a Benson, but what I could do is drive Dad's truck. Anything was better than staying there any longer. The keys were in the pickup, so I started it and let the engine run for a minute or two while I could still hear Dad arguing with the boys near the arena. I would run away from home, away from Dad, and away from all my misery. I slipped the transmission into first gear and drove to the gate that opened onto Post Oak Road. I turned right and instinctively headed for Highway 90 and towards Mimmy and Grandpa's. The consequences of what I was doing were profound. I had no plan, no idea what I might do, no clue where I might live, or even if I could live. This was likely to be the worst day in my young life.

The traffic was always heavy on Saturdays in Houston, so I just stayed in my lane and let the truck take me where it wanted to go. It felt like driving a mechanical Ouija Board, my fingers on the bottom part of the steering wheel, and some invisible spirit controlling the top. My head was spinning, but I realized I was on Highway 90, and the spirit was pushing us toward my grandparents. If Dad came looking for me, and he would certainly do that, Mimmy's and Grandpa's home would be the first place he would check. I was sure he had discovered by now that I was missing and that I had taken his truck. I was also sure he would not be coming for me right now. His rodeo would keep him in Houston at least until tomorrow, so that I would be safe for now.

Somewhere about halfway between Houston and Luling, I began to think I was making a huge mistake. I was a kid. I had no money. It wasn't safe to be at my grandparents' house. That would be the first place he would look. Even if I did go there, it would put Mimmy and Grandpa in a bad position. Dad could target them for harboring me, particularly if they lied to protect me. Then again, he might take out his anger on my mom or on my brother, and if he did, it would be my fault. I considered all these possible outcomes, but the invisible hands on the steering wheel forced me to continue putting distance between me and the monster that was sure to be hunting me down by tomorrow.

I don't know how long it took, but eventually I saw the city limit sign for Columbus and noticed that the gas gauge was telling me the tank was half empty. I didn't have a penny in my pocket and realized that I might not be able to make it all the way to my grandparents. I pulled off the highway, parked under the bridge that crossed the Colorado River, and turned off the engine. I needed to think. The decision to impulsively steal my father's pickup and run away was sinking in. How stupid could I be? One way or another, he was going to find me, and the outcome would be horrible, I mean, really very bad. But the deed was done now, and the only question was how to mitigate my punishment.

I watched the river flowing under the bridge and wished I could drift away like a floating log. It was made clear that my status as a son of James Albert Benson had been severed and that I was unworthy of being part of the Benson family again. Bensons would not, could not run away from a fight, the very definition of cowardice.

My thoughts wouldn't settle on any plan. Maybe I should leave the truck and hitchhike the rest of the way to Dewville. Maybe I should start hitchhiking and go all the way to California. Maybe I should use the remaining fuel in the truck to take it back to Houston, leave it at the arena, and then start hitchhiking somewhere. Another big log floated under the bridge. I felt like one of those lizards that could change colors from green to gray, to black, to pale white, depending on how it felt. I felt like a little dog that had just lost a dogfight. I oscillated between all these thoughts and all those emotions. I sat under the bridge until almost dark and decided to go back to Houston. There was a vanishingly slight chance that Dad was so busy with the rodeo that he had not discovered that I was gone in his truck. It would be dark when I got back, and Dad would be busy herding the bucking stock into the chutes and cursing everybody who did not meet his standards of efficiency. That would give me a chance to park the truck and get away. The show must go on.

CHAPTER 17

Hiding in the Hayloft

As I pulled off Post Oak Road and into the arena, I could see the bright lights mounted on multiple tall creosote poles illuminating the whole area. I could hear the announcer's voice issuing loud and clear from enormous metal horns, praising a well-completed eight-second bull ride. Since bull riding was always the final event, I had to move fast before my father was free to start looking for me. I was sure he had discovered the missing truck by now and was planning how to chase me down. He'd likely guess that the best place to look would be my grandparents' house, and he would put Mom and my brothers in her car and start driving west. But now, he would discover I'd returned the truck, with an almost empty tank, and he'd start searching for me around the property. I needed to hide and devise a plan to escape.

I considered taking one of the horses and hiding out somewhere in Memorial Park, but I decided against the idea. If Dad discovered a missing horse, he'd know I was somewhere nearby, and probably off in the park's pine trees. I needed more time to plan my escape. I was in the horse barn and thought about the hay loft above the stalls as a possible place to hide and think. I settled on the hay loft as the best place to evade his wrath. He would not be likely to climb up there and look for me, and even if he did, I could bury myself in the scattered hay and be well hidden. I was utterly exhausted and needed some rest, maybe even more than I needed to make a plan.

I climbed the ladder and found a suitable pile of loose, sweet-smelling alfalfa. I always loved the smell of alfalfa, as did all the horses. I crawled under the hay, gathered an armful, and made myself a sort of pillow. It was still cold in the barn, but the winter wind had died down. I tried to plot my next move, but exhaustion won out, and I was asleep in minutes.

I woke to the sound of voices. It was the voices of two teenage boys, just below me. They were Henry and Edward Withers, who often arrived early on Sunday mornings, for a ride along the many winding trails that weaved through the park. They were talking about one of their friends, Forrest Peterson, another teenage boy from the River Oaks community. So far, I had not met Forrest, but I knew his parents were determined for him to pursue a career as a concert pianist. He spent hours every day practicing, while Ed and Henry were free to have fun.

Almost all the customers who boarded blue-blooded horses at our stable were part of the River Oaks wealthy neighborhood just west of Memorial Park. I knew the boys and were sort of friends with them, although they didn't judge me as one of their class of humans.

In the 1960s, River Oaks was Houston's most prestigious enclave, a lush, meticulously planned neighborhood between Buffalo Bayou and the city's growing urban core. Grand mansions lined curving boulevards, set back behind manicured lawns and shaded by towering live oaks. The area exuded quiet affluence, with residents often belonging to Houston's old-money families, oil executives, and civic leaders. Social life revolved around the River Oaks Country Club, where tennis whites and garden parties were as much a part of the landscape as the azaleas that bloomed each spring. I could not imagine how life might be if I were born into such wealth and privilege.

After waking, there was no evidence that Dad was anywhere around. He had probably gone to Dewville to ensure I wasn't there and that my grandparents hadn't lied about my presence. He would have forced Mom to call and wake them in the night to see if I had arrived, but he would not have trusted them when they said no. I could not be sure of these assumptions, so I remained fearful.

I decided to make my presence known to Henry and Ed, and I climbed down the ladder to the vast, central area between the two long rows of stalls. The boys saw me right away.

"Hey, Bobby. What were you doing up there?" Henry said.

"My dad is after me for what I did yesterday, and you know how mean he is. He hates me, and if he finds me, he might kill me. I'm going to hitchhike somewhere away from Houston and never come back."

"Damn, are you sure that's what you want to do? How are you planning to live? Which way are you going?"

I knew that both Ed and Henry understood what a violent temper Dad had and how brutal he could be. A few weeks before, Dad caught me, Ed, Henry, and two other boys smoking cigarettes out in the woods near the stables. It was dark, and we didn't notice him sneaking up on us. We were horsing around and playing like adults, enjoying our forbidden, smoky pleasures.

Dad had worked his way behind me and raged out of the dark and grabbed me by the arm before I knew what had happened. The other boys were horrified and watched as Dad whipped me with a folded length of water hose.

"Man, Bobby. What the hell did you do, anyway?"

"He wanted me to fight some drunk guys that were working for him, and I was afraid to do it. My ankle was hurting, and they were almost grown-ups. They would have beaten me up badly. He yelled for me to get out of this sight and said he would deal with me later. I stole his pickup and drove almost to San Antonio. I brought it back last night before the rodeo was over and hid in the hay up there."

Then Henry floored me with an unbelievable suggestion, "You know what I think? We should all go over and talk to our father to see what he thinks is best. Your dad respects him, being a doctor and all, so maybe he can have a talk with him and fix things."

I almost couldn't believe what Henry was saying. Even though I'd known Henry and Ed for months, I'd never been invited to their house. I was dirty from working the day before and wrinkled from sleeping in the hay, and I had only spoken to Dr. Withers one time at his office, where he used stitches to close up my finger I'd pinched in a tire-changing machine. The machine had ripped off the tip of my left index finger, completely removing the nail with it. But I was certain that Dad would not find me at Dr. Wither's house, and pretty sure that even if he did discover where I was, he wouldn't be bold enough to come there to get me.

I took no time to decide that Henry's plan was a good one, and we loaded into Ed's pickup and drove to their stately house on Ella Lea Lane. It was still early Sunday morning, and Dr. Withers and his wife, Frances, were in the kitchen having breakfast. Of course, they were surprised to see me with their sons. Henry explained what I'd told him about my experience and what I'd suggested I was planning to do – run away to some place where my father could never find me again.

After listening to what had happened, Dr. Withers calmly decided, "Well, son, I think you should stay here with us until I can speak with your father. I could drive over to the stables and see if he is there, but I think we should let him stew about this for today, let him calm down, and I'll talk to him tomorrow if I can find him. He needs to be made to understand that his behavior toward you is not only wrong but also unlawful. It is part of my duty as a physician to recognize abuse and report it if necessary. Things are going to be made right."

Of course, I agreed with this plan. Ed took me upstairs and to a guest bedroom. The bedroom had its own bathroom and even a shower.

"Why don't you get yourself cleaned up in here. There's a bathrobe hanging on the door and fresh towels in the closet. I'll have Gloria (their maid) come up and collect your clothes, wash them, and dry them. In the meantime, just come on down in your robe, and breakfast will be ready. How do eggs and pancakes sound?"

All this was happening so fast, my head was spinning. I looked around the rooms and found it to be like no surroundings I'd ever seen. There was a big, fluffy-looking four-poster bed, a deep pile carpet, and a large window overlooking the manicured backyard. There was a pool in the shape of a watery kidney, and three expensive-looking bikes parked by the garden gate. Old live oak trees were spreading their massive, leafy branches to shade the yard, and scrubs lined the cedar picket fence. A curving flagstone path led to the pool, which was embedded in a large, rectangular, matching flagstone patio. The whole place, the front lawn, the stately two-story house, and the sculptured backyard squeezed me into a Hollywood-like fairyland.

I had finished my shower, gone downstairs, and had some scrambled eggs, two strips of bacon, and a stack of pancakes. My tattered jeans and striped shirt were washed, dried, and ready. Henry suggested that we take bicycles and ride over to the Mecoms' home to retrieve a shawl that Henry's Mom had forgotten after a party they'd attended there last night. I'd never heard of the Mecom family, or their across-the-street neighbors, but they were soon to play a crucial role in my future.

It was still pretty cold outside, and Henry had to lend me a coat, but bike ride to the Mecoms' house was surreal, like living a dream. I was so distracted that I didn't even notice the pain in my ankle. The pavement was smooth, unlike the potholed streets in Pasadena. There was no traffic, just Henry and me and the quiet hum of our tires. The air smelled like honeysuckles. These houses! They were like castles, with enormous white columns and lawns trimmed like they had their own barbers. Fancy lanterns were flickering even in daylight. I envisioned that the folks inside were sipping coffee from China cups and reading Sunday papers as thick as bricks. I wondered what it might be like to live in one of those homes. These mansions were even bigger than Henry's home, which was like a smaller mansion. Did the residents even notice the trees or the way the morning sunlight reflected from the leaves just perfectly? Were they so wealthy that they couldn't even see what they had?

It all filled my senses. I imagine parties there, with music and girls in dresses that swish when they walk. Probably some oil tycoon owned that one. Or maybe another doctor, but surely someone who never had to mow his own lawn. I didn't belong here, not really, but riding through it made me feel like I did. It felt like I was part of the story, even if just for a few blocks. We took the next curve. I let the breeze hit my face and marveled. Somehow, my fear of Dad and all the uncertainty about what was happening to me had faded. I felt safe.

But I had not seen what "real" wealth meant until we turned onto Lazy Lane. Mr. John Mecom lived on Lazy Lane, and his family's home was grander than the Withers' house; the jaw-dropping mansion across the street was unbelievable. Mr. and Mrs. Harry C. Hanszen owned it and were good neighbors with the Mecoms across the street. Later, I got into big trouble at that mansion, and the problem was a negative turning point in my life. There was no Google back then, so I only knew what people told me about the Hanszen home. I was told that it was a French chateau that had been torn down (stone by stone), transported to River Oaks, and reassembled. That turned out not to be true, but here is what Google says about the structure today.

The Hanszen family mansion in Houston's River Oaks was a storied architectural gem, inspired by Normandy, steeped in history, and once considered one of the most picturesque homes ever designed by Houston's legendary architect, John F. Staub.

Despite its legacy, the mansion was demolished in July 2017 by a new owner, sparking outrage among preservationists and heartbreak for some of the River Oaks residents. The teardown marked the loss of a cultural and architectural landmark, one that had quietly anchored River Oaks for nearly nine decades.

With the Hanszen mansion behind my right shoulder, Henry and I parked our bikes at one of the Mecoms' back doors and just walked in. One of the maids, Annie, was in the expansive kitchen. She recognized Henry, but she squinted toward me with a disapproving look. She was pretty sure the grungy-looking, skinny, and poorly dressed kid trailing behind Henry didn't belong in the house.

"Lord boy. You're here early, Master Withers. Who's your friend?"

"He's Bobby. Father is helping him out with a family problem."

She squinted at me again, as Henry explained, "Mom left her shawl here last night after the party. She wanted me to come over and get it?"

"Well, I saw it out there by the pool and brought it in. Didn't know whose it was, though. I hung it in the coat closet, but it's likely still damp. You know where it is, just go get it. But the folks is not up yet, so be real quiet."

Henry left me standing in the kitchen and started down the hall. I wasn't sure whether I should follow him or not. He didn't say anything or motion for me to follow, so I just stood there while Annie examined me.

"Lord boy, where you been living. You look like a bum. Don't your momma cut your shaggy hair sometimes?"

I squirmed a bit, "Yes, ma'am, but she's been busy lately."

I was cautioned later that I shouldn't say Yes, ma'am to Annie, since she was a black woman. I could remember when blacks were required to ride buses only in the back seat.

"Well, don't you allow Mr. Mecom to see you like that. He's upstairs, but you worry, he won't be coming down this early."

I said "Yes, ma'am again," and was glad when Henry returned with the shawl. That maid lady stared straight into my eyes, and I didn't see her blink once. It was really strange, but when she looked at me, I noticed my ankle was hurting.

Henry and I rode back to the Withers' house, taking a different route. All the homes I saw were larger than typical Pasadena houses and had at least two stories. The rest of the day, Ed, Henry, and I just played around doing all sorts of things. We played basketball out by the pool, Monopoly up in Ed's room, and talked a little about horses, because they had missed their Sunday morning ride. It was fun, but I was faking it somewhat. I felt like the Earth would be annihilated by a giant asteroid tomorrow, and no one knew that but me.

CHAPTER 18

WB5CXN

It is challenging to understand the minds of other humans. I've heard others say that they can do it, and maybe they can. I suspect that genuine empathy is impossible. Personally, I can learn the behavior of my fellow humans and develop a sense of what motivates them. In fact, I consider myself exceptionally good at this task. But the one model in which I have some confidence is myself. I am like a two-circle Venn diagram, and this diagram is presented differently for each person that I know, for each person I love, and for each person I have any level of contact with. For the people closest to me, the circles overlap significantly. For the lady who bags my groceries at HEB, there is almost zero common area. There is no other human or animal in the Universe who has the privilege to know all about my private circle. I am sure that that is the same for every sentient being. There must be a narrow private sliver of everyone's circle where some things that you know, some things that you did, and some things that you've lied about are stored away. The one commonality among humans is that they all, consciously or unconsciously, engage in lying. Lying may be the single most defining human trait.

I don't remember where or how I heard that the only thing humans can know for sure is that they (their consciousness) exist. In other words, "I am." Maybe I am the only sentient being in the Universe, feeling lonely, and mentally fabricating the entire Universe to amuse myself and ward off loneliness.

I do not believe there is an afterlife, or awareness after death. But rationally, I have no proof of this. There could be an afterlife, maybe heaven, and hell, but I assess that the chance of these things is so vanishingly small that I don't waste my precious life-seconds worrying about them. Let's get grounded again. Here's where we are in my story.

I am a young, uneducated bricklayer who sees a dim future if I don't make some monumental life change. I am presently working on a job in San Marcos, Texas (a college town), and have felt a tiny crack in my Universe. The thought that I might be able to change my life by attending college has infected my mind and worked its way into my bloodstream. A little thought like that can be as debilitating as contracting mental malaria. One single-celled "Plasmodium" has hidden itself in my liver, busted out, and is now infecting my mental red blood cells. Like real malaria, the disease has divided and is now multiplying exponentially in my mind.

Okay, now, what do I know about malaria? I contracted malaria in the 1940s as a small child. At least that's what my mother wrote in my Baby Book, along with assorted other child trivia (when I cut my first tooth, uttered my first word, fell and cut my lip, etc.). It is difficult to believe, but malaria back then was a severe and omnipresent disease in Texas. In fact, it was still considered a significant public health threat across the southern United States at that time. Entire communities witnessed mosquito eradication campaigns, which included trucks spraying pesticides (DDT) down city streets. Public health efforts included draining swamps, installing window screens, and eliminating standing water.

My kid friends and I are lucky to be alive, if they still are. The DDT spray truck used to come down our street around dark each day. One of our games was to run behind the slow-moving truck and play in the whitish, oily vapor billowing from pipes on each side. It had a mild, musty chemical smell, and Mom forced us to take an after-truck bath to try to wash it off. I know that about malaria.

I have several quirky hang-ups. Getting anything oily on my body drives me crazy. I've travelled to Belize and Peru a few times and have always been cautioned to put on bug repellent to keep the malaria-laden mosquitoes from biting me. I refused to do it. Bug spray is oily, and I hate oily, so guess what? A mosquito sucked my blood, and I got malaria for a second time. I don't remember my childhood malaria, but I'll never forget my adult malaria. It came close to killing me.

Don't you hate it when your narrator drifts off topic, as I just did? Back to San Marcos, then. I have been offered the opportunity to experience college life (vicariously) by moving in with some students who view me as a means to help cover their modest monthly rent. I have decided to spend a weekend driving a circuitous route from San Marcos to my grandparents' place near Dewville, Texas. Aside from stopping now and then to observe and count wild birds (one of my diversions), I've been telling you about my early life. I have arrived at my grandparents' home, and I cherish the moments I spend with them. But before returning to San Marcos, I need to explain what happened when the Withers family decided to save me from my mostly brutal, but sometimes fun, unstable father.

I stayed all that Sunday with the Withers. That night, they put me in the guest room. I didn't know guest bedrooms could look like this one. I mean, I'd seen fancy stuff on television and in movies, but this was different. This was real. I was in it.

The carpet was so thick my bare feet sank into it like mashed potatoes. Cream-colored, soft as a dog's ear, and not a single stain. Not even a scuff. The walls were painted pale gray that looked silver when the light hit them just right, and the ceiling had a chandelier, not the crystal kind, but something modern, like a rocket ship made of glass and gold.

There was a bed in the middle, king-sized, with a headboard that looked like it belonged to a museum: tufted velvet, deep blue like the sky before a storm. The pillows were stacked like a hotel ad, with big square ones in the back, followed by medium ones, and then tiny decorative ones that didn't make sense but looked expensive. I didn't touch them. I felt like I'd mess them up just by breathing too hard.

To the left was a dresser that surely cost more than Dad's truck. Sleek, shiny wood with brass handles, and on top of it sat a bowl of green apples that looked fake but weren't. I checked. They smelled like apples. Next to that was a lamp with a marble base and a shade so clean it glowed.

The window was huge, floor to ceiling, and it looked out over a backyard that could've been a golf course. Manicured hedges and a stone path that curved like it had been designed by someone who studied curves for a living. I could see the tops of oak trees swaying like they were in a movie. River Oaks trees. Rich people trees.

I scanned the room for a while, just taking it in. I felt like I was trespassing, even though they'd invited me. It was like borrowing someone else's life for a night. I didn't sleep much. Not because the bed wasn't comfortable, it was, insanely so, but because I kept thinking about how different everything felt. How quiet it was. How clean.

How far away it was from the world I knew. I wasn't jealous. Just... curious. Like I'd stepped into another universe where even the air smelled like money.

The next morning, after breakfast, Henry invited me to join him for a visit to one of his friends who lived in the Spring Branch area of Houston. Henry, who was also 16 years old, had a driver's license and was permitted to drive his mother's Cadillac. His friend, whose name I don't remember (I think it was William), was an amateur radio operator, or "ham," as Henry called him.

Spring Branch was not as ritzy as River Oaks, but still a lot more upscale than Pasadena. The houses there were modest, mostly single-story ranch-style homes built for middle-class families, with brick façades and large yards. We pulled up to Henry's friend's house, and the first thing I saw was this monster radio tower looming thirty feet high from somewhere behind the house – so big that the tips of an H-looking structure were sagging toward the ground. I suspected it was an antenna of massive proportions. It is a spherical contraction below the antenna that was guessed to be a motor. I asked Henry what the thing was.

"What is that thing hanging from that tower in the backyard?"

"That's Willy's 80-meter Yagi antenna. He's got a control box in the shack that lets him point it to different places in the world. Right now, it's pointed toward the south, so he's been working South America, probably Peru or Brazil."

As I've mentioned, I had been fascinated by anything bordering on science since I was a child. I didn't understand a word that Henry told me about the antenna, so I hit him with a barrage of questions. What is Yagi? What do you mean by working in South America? Is that round thing at the top of the tower a motor, and does it make the big H turn fast or slowly? Why doesn't that black wire going up the tower get all twisted up and twisted around the tower?

"Let's go in and get Willy to show you all about it."

We knocked on the door, and Willy let us in.

I liked Willy right away. He was a strange-looking boy. I'd played around with a crystal set, a simple kind of radio receiver a kid could make from cheap parts from his local hobby shop, nothing more than a chunk of Germanium embedded in a lead cylinder, a cat whisker (a little steel wire that touched the crystal, a resistor, and some copper wire. The resistor was a small, rigid cylinder with three colored bands painted on the end. The colored bands indicated the resistance of the resistor.

Willy looked like the kind of boy who could recite resistor color codes faster than most kids could name their favorite baseball players. His bedroom was a shrine to vacuum tubes, soldering irons, and the soft hum of his shortwave radio rig, patched together from surplus Army parts and hobby shop scraps. He wore thick glasses, his shirt perpetually untucked, and spoke in bursts of technical jargon that bewildered me. While others dropped nickels in jukeboxes and frequented drive-ins,

Willy chased signals bouncing off the ionosphere, logging distant voices from Moscow, Manila, Sidney, and places like that. I guessed he would be awkward in crowds but magnetic behind his microphone. Beneath the geeky exterior was a boy with boundless curiosity and a quiet belief that radio waves might be the threads that stitch the world together. I was smitten with all the wires and radios, and Morse code sending keys, and all the postcards pinned on his wall sent by other hams around the world with whom he had made contact. His whole bedroom, his shack, as all hams called the place where their equipment resided, was alive with dials, knobs, microphones, wires, and scratchy, popping sounds from a speaker. I was mesmerized. This seemed like magic.

Willy showed me how it worked. He pushed a lever on a little box, and I could see the big H in his backyard turn slowly to the right. He said he might be able to talk to someone in South America on 80 meters, the length of a single wavelength at the frequency on which he was transmitting.

Once the antenna was in position, he let go of the lever and pushed down a button on his microphone. He began talking in a coded language.

"CQ CQ CQ pause CQ CQ CQ, pause, this is WB5CXN, WB5CXN, South Texas calling CQ CQ CQ."

The letters WB5CXN were Willy's station's call letters, issued by the Federal Communications Commission. He stopped and listened to see if someone might answer his call. No answer.

He repeated the sequence, "CQ CQ CQ pause CQ CQ CQ, pause, this is WB5CXN, WB5CXN, Houston, Texas calling CQ CQ CQ," and listened again.

This time, I heard a scratchy voice speaking English, but with a heavy foreign accent.

"WB5CXN, here is HK3BHZ, HK3BHZ, Bogotá. Your signal is FIVE NINE, here in Columbia."

I was beyond excited. Here we three were, standing in Willy's bedroom, his shack, and talking to another person, in another country, who was in South America. My heart was racing! This had to be the most exciting thing I'd ever seen or heard. It fit perfectly into my still naïve concept of what "science" was. Right there and right then, I decided that I must learn how to do this and have a big antenna and a shack with radios, knobs, and microphones.

We stayed about an hour at Willy's house and then rode back to River Oaks, but Henry wanted to stop by Forrest Peterson's house to drop off something. Forrest did not have horses at our stable. I knew his name but hadn't met him. I also knew he spent a lot of time practicing the piano.

The Peterson house sat like a museum on a quiet, tree-lined street in River Oaks. It had white columns, manicured hedges, and a driveway so clean it looked polished. Inside, the air was cool and perfumed with something floral and expensive. A woman in pearls greeted me with a smile that didn't quite reach her eyes. "You must be Bobby," she said, glancing at my hands. I shoved them in my pockets, suddenly aware of the raggedness of my nails.

Forrest came in from the hallway, tall and lean, with a mop of dark hair and a posture that looked trained. He wore a crisp button-down and loafers that probably cost more than my father's truck.

"Hey," he said, casually, like we were equals.

I could intensely feel our class gap. I suppose that Henry wanted to show off his friend. He asked Forrest to play something for us.

We were ushered into the music room, which was, indeed, a room just for music. A Steinway grand piano dominated the space, its black lacquer gleaming under a chandelier. Forrest sat down like he'd done it a thousand times, fingers hovering above the keys.

"My parents want me to audition for Juilliard," he said, almost apologetically. "I'm not sure I'm good enough."

Then he played.

It wasn't just good, it was haunting. The notes floated like smoke, delicate and precise. I stood there, stunned, feeling the calluses on my own hands, the remaining ache from my twisted ankle, and the sudden realization that boys in River Oaks didn't dream of getting out of their daily lives. They were already somewhere most of us couldn't even imagine.

When he finished, Forrest looked up, expecting perhaps some praise. I nodded, unsure what to say. "That was... something else."

He smiled, and for a moment, the class divide blurred. But as his mother reappeared with sparkling water in crystal glasses, I remember how often I'd drunk from a garden hose.

When we got back to Henry's house, Dr. Withers was still at his clinic, and neither Henry's mother nor his brother was home either. So, we spent the rest of the day with Henry trying to teach me how to play Chess. I picked up the ways the pieces moved on the board, but I was clueless about the deeper strategies of the game. Maybe I was distracted by the thoughts that kept creeping back into my head. I wondered what Dad was doing. Had he gone to Dewville and been mad at my grandparents because he thought they were lying about whether they knew my location or not. Was he taking his anger out on Mom, or maybe on Johnny? I got to thinking about my girlfriend in Pasadena, Shirley, that I've mentioned before. I missed her desperately and was attached to her with an intensity only possible in the throes of teenage infatuation.

It was Christmas holidays, and I knew Shirley would be at home, unless she was out with some of her girlfriends. After Henry and I got tired of Chess, I asked him if I could call her on their phone, and he said, "Of course."

I talked to Shirley for about an hour on the phone, but I wouldn't tell her where I was, because I was sure the McGlothlins were the second place Dad would be looking for me if I wasn't in Dewville. Passiveness was not in me.

If I did something or felt something, I did it or felt it with passion. I was passionate about Shirley. I loved her and knew she was in love with me. Even if the grown-up would call it "puppy love." There was even a popular song about it when I was a kid, "And They Called It Puppy Love," by Paul Anka. Shirley and I thought that song was written for us. Anka didn't just perform, he connected, channeling youthful longing into timeless melodies.

Finally, Dr. Wither's car pulled into the driveway. What he was about to tell me would change my life forever.

CHAPTER 19

Gray Eyebrows

As you will imagine, my two days with Henry, Ed, and my brief visit with their dad, Dr. Withers, and his wife, Frances, were marked by both a sense of wonder and a nagging feeling of dread. The River Oaks life seemed like a Hollywood movie, but even better than the cheap kid movies my brother and I used to see on Saturday afternoons at the Hi-Ho Theater on South Presa Street when we lived near the state hospital. The difference between Pasadena people and River Oaks people was like the difference between Old Yeller and East of Eden. I couldn't push away the feeling that the devil was taunting me by showing me this different way of life that I could certainly never have for myself.

I'd been allowed to ride around on a Bianchi bicycle, known for its prestigious brand, celeste green frame, and its Giro d'Italia pedigree. My real life was a rusted-out Sears and Roebuck sidewalk bike with a steel frame, single speed, terrible styling, leaky tires, and sold through catalogs.

Dr. Withers sent the maid upstairs to fetch me for a talk about what was to be done. It was to be a family meeting in the living room downstairs. Henry and Ed were called to join, and Mrs. Withers would be there, too. My hands got sweaty, and my heart rate went up. I could only think of options like Dad had promised that I would not be punished for what I had done, and that he had assured Dr. Withers he would moderate his corporal punishments in the future. I knew he was a good liar, and as soon as Dad got me home, I was in for the beating of my life. This family meeting was to inform me of the arrangements my father had made with him. I would be loaded into Dr. Withers's Cadillac and driven to the rodeo arena and turned over to my smiling father.

When everyone was seated in the living room, Dr. Withers began. "Well, Bobby, this afternoon, I had an opportunity to speak with your father and mother. I asked them to come by my office, and we had a good chat. I made it clear to your father that his frequent abusive punishments would have to stop; that I was bound by law to report such abuse to the Department of Public Welfare. That department could choose to report his behavior to the Harris County Prosecutor's office for possible criminal charges. I think that got your father's attention."

I could see where this was going. "Sir, please don't take me back to my father. He is lying. He can't help himself. It was the war that made him crazy, and when he gets mad, he won't care about what he promised you. Just let me go, and I can hitchhike to my grandpa's house and stay there until I figure out what to do."

"Well, son, I understand your fear, but I don't think that is a satisfactory solution. You shouldn't be hitchhiking, and your grandparents don't need the burden of protecting you and sending you to school. In a normal situation, those tasks are relegated to your parents."

I didn't know what relegated meant, but I didn't like the sound of the word. "I can't stay there anymore. He might really kill me for taking his truck like I did. Thank you for your help, but I can't go home." I looked around the room and knew that this was the last time I'd ever again see a room like this. I could see the shock on Henry's and Ed's faces. Even Mrs. Withers seems disturbed.

Dr. Withers spoke up, "You are not going back to live with your parent for now. I have not told our boys, but I did discuss a plan with my wife," he looked toward Mrs. Withers, "and we've decided that you will stay with us and finish out the school year here with Ed and Henry."

My thoughts were running wild. This could not be true. It felt like a cruel joke. My hands got sweaty. I even felt like I would cry. I felt dizzy and sat back on the sofa.

"I can see that this is a surprise for you. But you and my boys get along well, so it will be no inconvenience at all. If you like the school, and I know you will, then we can think about making this arrangement permanent. I mean, having you stay with us until you graduate."

I began to think Dr. Withers meant what he was saying. "Will I still have to work for my dad at the riding stables?"

"No, you could be there with Henry and Ed, only if you wanted to. Your parents agreed that they would not interfere. This will give you a fighting chance to recover from your trauma (another word I didn't understand) and make something of yourself."

I'm unable to express in words how I felt, but relief is undoubtedly one of them, exhilaration another. Then, I thought about Shirley. I loved her, and even if it was teenage love, the thought of not seeing her burned a hole in my heart. It was 18 miles from River Oaks to Pasadena, right through the middle of downtown Houston. I knew it was the wrong thing to bring up at this moment, but I said, "I have a girlfriend in Pasadena. I don't want to lose her."

Dr. Withers looked disappointed, but said, "Well, accepting our offer requires you to make some choices. This opportunity could change the direction of your life. You may call her on the telephone, and you could ask Henry to drive you there from time to time, but that's not something we could accommodate every day. You'll have to make some new friends at school here."

Dr. Withers had bushy gray eyebrows, and for some reason, those now became my focus point as I watched his face. It was strange to me that I had not considered losing Shirley when I was contemplating running away from home to some place like California. River Oaks was a lot closer to her than being some place like San Francisco. Something deep inside spoke to me and insisted I be rational, even though rationality was not in my character. Again, I looked around the room. I saw them waiting for me to speak. There was Henry, Ed, Mrs. Withers, and Dr. Withers waiting. "I think I want to try. Mrs. Withers got up and hugged me. I was now a River Oaks kid, but maybe not as long as I hoped I'd be a River Oaks kid. I sat there in that fine living room, surrounded by people who had no reason to care about what happened to me, and tried to understand why they did. I couldn't. But I didn't need to understand it. I just needed to not ruin it.

CHAPTER 20

Purple Pups

I was too young to understand or care about the legal requirements that allowed the Withers to take me in and enroll me in a new school. I guessed that it was all informal. Powerful families were rarely questioned about their decisions, such as taking in a blue-collar kid and enrolling him in school. I was soon enrolled at Sidney Lanier Junior High School in the Montrose area of Houston. I laughed at Lanier's mascot's name. We were referred to as the Purple Pups. Kids from my old junior high back in Pasadena were the Bulldogs. Considering the manors and behavioral differences between the schools, the mascots and nicknames seemed appropriate.

In the 1960s, schools were still segregated, and Lanier was an all-white school. The building had the classic American schoolhouse look, with its red brick architecture, arched windows, and decorative stonework. It almost looked like a small college. Unlike the working-class kids in Pasadena, the students at Lanier wore fancier clothes. Jeans were still considered too casual for school, so boys wore pressed trousers with belts. Penny loafers and saddle shoes with plain socks were standard, so Mrs. Withers took me to Sakowitz Department Store to get me all decked out with what Sidney Lanier expected. My family, of course, had never shopped there. I was "gobsmacked" (not my teenage word) by the place. It was five stories high and looked like an art museum. After shopping, where Mrs. Withers selected all my new fashions, we had lunch at the Sky Terrace restaurant at the top of the building.

I could go on and on about how vastly different my life had become and what a different trajectory it had taken, but the distance was too far. My behavior ultimately strained my relationship with the Withers and jeopardized, no shattered, what could have been my ticket to an upper-class life.

As could have been anticipated, I didn't fit well at Sidney Lanier. The fancy Sakowitz dress could not hide the way I spoke or conceal my class differences. In Pasadena, students were destined for blue-collar jobs (oil refinery work, bricklaying, and similar occupations), not college futures and white-collar desk jobs. Even though my classmates at Lanier were cordial enough, I could feel that cavernous gap between them and me. As time went on, a great tension developed, compounded by my almost severed relationship with Shirley. I did get to call her sometimes, and only once did Henry agree to take me to see her in Pasadena. It was very awkward because we could not be alone with Henry there. I'm sure Shirley missed me as much as I missed her, at least I thought so. I have always been prone to emotional extremes, and being ostracized at Lanier acutely dramatized my longing for her affection and companionship. I turned 17 in February of that year, and 17 is probably the peak of teenage hormone derangement.

Here's how it all went down. Earlier, I mentioned the close relationship between the Withers and Mecom families and their status as pillars of the River Oaks community. Henry, Ed, and I often visited the Mecom home. Now, the Mecoms and the Hanszens (who owned the mansion across the street) were good friends, so we also spent a good deal of time at the mansion and on its vast manicured grounds. There is an odd circuitous connection between John Hanszen (the philanthropist), who funded several building projects at Rice University, and my family's story. Many Rice students affectionately refer to themselves as Hanszenites. That odd connection is that my brother John attended graduate school at Rice and was awarded a Ph.D. in Space Physics and Astronomy. I never heard him say he was a Hanszenite.

The mansion's grounds were so extensive that the Hanszens had a Motobécane moped, which was used to ride around the place. I loved driving that thing. I also loved Shirley, and that's where my trouble started, and salvation ended.

One fine spring day, I was riding the moped and missing Shirley, and you can guess what a stupid thing I decided to do. No one was watching me, and Pasadena was calling me, so I drove that Motobécane right out of the gate onto Lazy Lane and headed for Pasadena and Shirley. Thinking back now, I can't imagine how stupid this impulse was. If I stayed off the Gulf Freeway, the distance would be around 25 miles through the tangled arteries of Houston traffic. And this, on a machine that topped out at 30 mph downhill with a tailwind. No helmet. No license plate. No plan. Just raw emotions.

I kicked it to full speed, and the little two-stroke engine was coughing like a chain-smoking squirrel. The exhaust smelled like burnt oil and rebellion. As I rolled past the mansions of River Oaks, I felt like a mosquito in a Rolls-Royce showroom. Drivers in Mercedes and Cadillacs gave me the kind of look reserved for stray dogs and bad ideas.

Then came the real test: merging onto Allen Parkway. The moped wheezed as I tried to keep pace with traffic. By the time I hit the East End, my legs were numb from vibration, and I was pretty sure the engine was losing compression. I pulled over at a 7-Eleven store and let the motor cool off for a while.

A man in a cowboy hat looked at the moped and asked, "You rode that here?"

I nodded. He shook his head slowly. "Son, that's either brave or stupid, and probably both."

I finally limped into Pasadena just before noon, smelling like gasoline and triumph. The moped was rattling like a tin can, and I was one pothole away from disaster. But I made it to Shirley's house, somehow. Looking back, it was a terrible idea. Dangerous, illegal, and wildly irresponsible. But it was also the kind of story that sticks to your ribs. The kind that reminds you what it felt like to be young, invincible, and just dumb enough to live through the experience.

Of course, as soon as the Hanszens discovered the missing moped and knew who'd been riding it, they called the police. It didn't take long for them to interview Mrs. Withers, who was at home, and predict where I had gone. Pasadena was a separate municipality with its own police department, so the Houston officers contacted the Pasadena officers, and before long, they arrived at Shirley's house.

I was terrified and was pretty sure that they would be taking me to jail. They handcuffed me and sat me in the police car. It smelled like fresh vomit mixed with vodka, and the stench was making me sick, too. I stayed there for what seemed like hours, while they talked to Shirley and her mother and wrote down a lot of stuff on clipboards. Another police car came to the house and parked behind the one I was in. I could hear them discussing my case on the police radio. They would say something, and then there would be a kerr-chunking sound. I'd listen to the radio blaring. I couldn't understand much of it. A pickup arrived with police markings, loaded the moped, and carried it away.

Then another car arrived! A car that I recognized. It was my mother in her old Chevrolet. What was she doing here? I had not seen her since I moved in with the Withers. I had only spoken with her on the phone a couple of times. Now the police were interviewing my mother. I had been ordered to remain in the police car, and now I needed to use the restroom. No one was talking to me at all. I hoped I could hold it.

Finally, an officer and my tearful mother came to present me with the verdict. The officer grabbed me by the arm and, rather roughly, pulled me from the back seat. The last thing I saw in the police car was the perforated, cage-like partition between the back and front seats. It was like a miniature jail and reminded me of the cages I'd seen in the backs of the dogcatcher's pickup. Even if the dog had peed and pooped in those cages, it could not have smelled as bad as this police car. The policeman unlocked my handcuffs, and fresh blood flowed into my white fingers. I still needed to pee myself, but I was pretty good at holding it off.

"Okay, young man, here's how this is coming down. You know you screwed up by taking that moped and driving it all the way across town in heavy traffic, causing regular drivers to swerve around you and endanger themselves. That thing isn't licensed for public roads, anyway. We could put you in jail for that alone."

Mother continued her sobbing, and I tried to look up at the stern face of the officer. My vision was distracted by the falling of dead tallow tree leaves in a gust of wind. I wished the policeman would get to the point and let me face whatever punishment I had coming. I was sure it would not be belt-whipping. That horror would come from my father later. Most probably, I would be put in juvenile detention. I knew of some other boys from Pasadena who had been sent to juvenile detention, but I didn't know what they made you do in that kind of jail. I felt dizzy. The officer saw my attention drifting.

"Pay attention, boy. This needs to be a serious lesson. Do you understand?" I shook my head.

"We had our dispatcher speak with Mrs. Hanszen, who owns the moped. She called this matter into our station when she discovered the bike was missing. She knew it was you who took it. But she said that she did not want you punished for it. She said that they had given you their permission to drive it around their property, but not all the way across town. Since they did not explicitly instruct you to remain on the grounds, she feels it would be unfair to charge you for stealing. So, we are not charging you for that."

I was not exactly sure what this meant. I wasn't going to be charged with stealing, but he said I was breaking the law because the small motorcycle didn't have a road license.

"What I am going to do is issue a citation, and you will have to pay a fine. That infraction will go on record and stay there for the rest of your life.

I understand that you have been living with a family on the other side of town, is that right?"

"Yes, Sir."

"And that family is the Dr. Henry Withers family?"

"Yes, Sir."

He looked at my mother. "We are going to release you without charges to your mother here. She may take you home and choose whatever punishment she or your father deems appropriate. Your mom says that she's received a call from the Withers, and she'll let you know what they have to say about your behavior."

He shook my mother's hand and sternly admonished me, "You'd better let this break we're giving you be a lesson, boy." He walked away, back to his car, and its flashing red lights.

It wasn't far from Shirley's house to our house, and Mom was silent as we drove. She had stopped her sobbing.

"Bobby, you are not going to like this, but I need to explain what Mrs. Withers said about what you did. First, she said that you were not fitting in well with the young people at Lanier and seemed unhappy most of the time. She mentioned that you were older than the other children in your class, and that evidently was part of your problem."

"But Mom, I'm not unhappy, I just miss seeing Shirley. That's why I did it."

"No mind about that, the important thing is that they don't think you belong there anymore. They don't want you back there because they trusted you, and you broke their trust. She said that your behavior embarrassed her and had stained their relationship with the Hanszens. She said that she would ask Ed to gather up all your things and bring them to you in Pasadena. Those things, of course, would include all the items, clothes, and school items they had purchased for you. Once that is done, she asks that you refrain from future contact with her family."

How could I have been so stupid? Looking back, it's hard to gauge how much that misjudgment affected my life. Certainly, screwing up with that stupid moped was a serious blunder. Had I stayed with Henry, Ed, and Dr. and Mrs. Withers, my future would have been different. I might have become a lawyer, an oil tycoon, or even a physician (of some kind that didn't have to see blood). It didn't turn out that way, so that's why you started learning about me on that construction site, on that scaffold, on that hot August day, in a college town on the edge of the Texas Hill Country, as a young man with slim prospects for a bright future.

That night, when Dad got home from work, I received the lashing I expected. It was exceptionally brutal. My butt and back were covered with crisscrossed, parallel, puffy bands of swollen tissue that looked like fossil leaves squashed between two plates of ancient slate. Mom tried to enroll me in school, but they didn't want me back. They suggested that I should be shipped off to the Army, so maybe some discipline would straighten me out. That's what my mom and dad did, but the Army didn't work. After several months of training and trying hard, the military decided I was too fucked up to make even a poor soldier, so they discharged me early. Honorably enough (compassion, really), so I would have full benefits under the GI Bill.

CHAPTER 21

The Mad Dog Story

Another strangeness I have is my sleep patterns. It is difficult for me to fall asleep in a silent room. I need the sound of a fan blower, or even better, a voice on the radio. I had none of that at my grandparents' home. If the sounds I heard from outside the house were frogs, hogs, or other wild animals, the sounds from inside were the ticking of wind-up clocks. My grandparents did not have dogs, but that night I heard the distant barking. That made me remember a scary story my grandmother told. A story about a dog and rabies.

My grandparents had always called it hydrophobia, and that word carried more weight than "rabies" ever did. A whisper of a "mad dog" sent families into a panic. Children who usually ran barefoot in the fields were herded indoors, doors bolted, windows latched. Men who feared neither rattlesnakes nor fever tightened their grips when they heard hydrophobia spoken. That fear seeped into me young, and the sight of that staggering dog pulled it out of hiding, sharp as ever.

With it came the old story I had heard since boyhood, the one my grandmother told with her voice low and her eyes fixed, as if she meant to scare the lesson into us. It was the story of her cousin Bettie, who died in the late 1880s. Whether every detail was accurate didn't matter. In our family, it carried the weight of the gospel.

Bettie lived with a bad hip, and it slowed her. One windy day, she was hanging sheets on the line when a hound wandered into the yard. Stray dogs were common in the open countryside, and she thought little of it until she saw the foam dripping from its mouth. Fear gripped her. She dropped her clothespins and turned for the porch. The dog lunged, seizing her skirt, teeth tearing fabric but missing flesh. She stumbled inside and slammed the door just as the hound crashed against it.

She sat shaking in her rocker, listening to the door rattle with the animal's weight, growls seeping through the cracks. Then her eyes found the shotgun that was propped up by the fireplace. She thought of her husband, out plowing, unaware of the danger. Her hands steadied. She rose, cocked the gun, and pushed the barrel through the screen. The hound's eyes rolled white. She pulled the trigger. The blast threw it back into the yard, and the struggle fell silent.

That could have been the end, but my grandmother always leaned forward then, her voice dropping lower. Bettie had escaped the dog's teeth, but not its sickness. The next day, she mended her torn skirt. She threaded her needle and, as women did then, snapped the thread with her teeth. No one told her the fabric was stained with the dog's saliva. With each bite, she drew the sickness into her mouth.

Weeks later, she grew restless. Her throat tightened. She gagged at the sight of water. Neighbors called it hydrophobia. They tried what cures they knew. Some spoke of the mad stone, said to be cut from the belly of a deer, placed on the wound to draw out poison. Others wanted to plunge her into cold water, hoping that shock would help. The doctor shook his head. There was no cure. Bettie's body arched and fought itself, foam on her lips, until the sickness burned her out. The family finally carried Aunt Bettie to the Northeast end of their farm, and tied her to an old oak tree. They were God-fearing folks and concluded that he had decided this would be her fate. It took two more days for her to die.

We never questioned the story. As children, we believed it the way we believed thunder brought rain. Bettie's death made rabies more than a word; it made it a monster waiting behind the next fence. A bark in the night could freeze us in place. The tale pressed itself into our bones, and it never left.

I once saw a dog limping along the roadside outside Prairie Lea, the fear came alive again. I kept the car steady and my eyes forward, but inside I was already braced for the worst. The dog staggered into the ditch, shadows swallowing it, but the memory remained.

I pulled off at a wide spot and let the engine idle down. Prairie Lea was never more than a handful of houses, a school, a store or two, with fields stretching out beyond. Cotton still grew nearby, though the gins had mainly closed. Even in stillness, the town carried its history on the weathered boards and leaning fences. My grandparents used to talk about the cotton gin, the blacksmith, and the feed store. Those buildings were gone, but the air still seemed heavy with the weight of old labor.

I sat in the car thinking of Bettie. I could see her hands on the cloth, the thread between her teeth, the sickness she never knew she was drawing in. I imagined her gagging at water, her family helpless, the neighbors whispering prayers that could not help her. My grandmother had warned us, and it worked. Decades later, a single sick dog could still pull me back into her story.

The older I grow, the more I see what those warnings really were. They were a form of love, heavy but honest. In our family, advice rarely comes as lectures. It came as short stories meant to stick. Bettie's story stuck harder than most. It told me that danger can hide in the plainest act, that you can be standing in your own yard, stitching your own clothes, and still invite death in without knowing it.

Some say country children grow tough. Maybe so, but toughness doesn't mean the absence of fear. It means living with fear close at hand and still moving forward. By the time I was ten, I could tell the sound of a rattlesnake's warning, the shape of a cottonmouth in water, the look of sickness in a dog's eyes. I learned to step carefully, to keep still when I wanted to run, and to listen when the old ones spoke.

CHAPTER 22

Cistern to the Creek

Late in the night, it began to rain. It came upon the darkness suddenly. One minute, I could still hear the distant dogs barking, and the next were the sounds of large raindrops, first on Aunt Eve's and Aunt Rachel's cistern just a short walk to the east. I could probably throw a rock from my grandparents' house to the old shack of a house in which my two aunts lived. They cooked on a wood-burning stove, and all the water they used ran off their tin roof and into their cistern. Their house was an archetypal clapboard 19th-century house, and that's the time in which they lived, too: the 19th century. The heavy drop upon their cistern revived memories of an experience my brother John and I had as children. Both of us could have died in a foolish thing we did with a cistern.

I can still picture its cold, metal exterior and the dark water that sloshed inside. I'll recount the tale, and you might find it hard to fathom just how reckless children can be. Ever since I can remember, taking risks has been second nature to me. The number of times I've brushed against disaster is countless, not because I chased after adrenaline, but simply because I never paused to consider the catastrophic consequences my actions might unleash. This penchant for danger took root early in my life and persisted for many years. Now, as I write these words in my eighties, that adventurous spirit remains, albeit tempered with a touch of caution. Let me share with you one particularly harebrained scheme from my life; a standout among countless other foolhardy escapades, that somehow, by sheer luck, I survived.

Our father frequently uprooted us, leading us to inhabit a series of dilapidated rental homes scattered across Texas. I vividly remember one instance when we lived near Richardson, Texas, in a house precariously perched on the high bank of a deep, dry creek bed. My brother John and I, as was our custom, roamed the area without any adult supervision. We were particularly drawn to a nearby small airfield, where we spent hours watching the Piper Cubs and other small aircraft as they ascended and touched down on the runway. Occasionally, we would muster the courage to plead for a ride in one of these fragile, fabric-covered planes. The thrill of sitting in those cramped cockpits, feeling the rumble of the engine before us, ignited our lifelong passion for flying. It's almost unthinkable now, a time when a stranger would willingly take a child, without any parental consent, for a joyride in a hand-cranked two-seater airplane. Today, such an act would likely lead to legal trouble. Back then, the freedom we children enjoyed and the way parents approached shielding their kids from potential dangers and unsavory individuals were worlds apart from current practices.

Have you ever seen a cistern? Picture a massive cylindrical container designed to store water, primarily rainwater collected from the roofs of old homesteads lacking a water well. These oversized "cans" were carefully positioned on a low platform beside the house, high enough to create a gentle trickle of water pressure indoors, yet low enough to capture the rain cascading from the roof into the top opening.

Our fascination with the small planes at Richardson Airport may have sparked our curiosity in crafting a precarious contraption ourselves. In the backyard of our Richardson house, an old rusting cistern lay on its side, abandoned and forgotten. It was about eight feet long, with a diameter roughly the same, its conical roof and one side corroded and eaten away by rust, while the rest remained surprisingly sturdy. The metal ribs of the cistern still held firm, like the skeletal remains of some ancient beast. Despite being just small children, John and I could manage to roll it around if we pushed hard enough on its side. We often did this, laughing and tumbling as the cumbersome cylinder wobbled across the grass, driven by nothing more than our childish whimsies. There was a moment when we thought about sending it careening down the hill into the dry creek below, but the thought of Dad's fury stopped us short.

Then came a day when a reckless idea sparked in our minds, one that could have ended in tragedy. I'm incredibly fortunate to be here recounting this tale. It's as if fate has flipped a coin fifty times and each time, it miraculously landed on heads.

Amidst the haphazard jumble of rusted cans, broken glass, and twisted metal that cluttered the ground around the dilapidated shed at the back of our house, a few treasures caught our eye. We spotted a ten-foot wooden pole, weathered and rough to the touch, and a rickety chair made of flimsy metal and cracking plastic. These were exactly what we needed to transform our old cistern toy into the cockpit of an imaginary airplane or spaceship.

With Dad's hammer, its handle worn smooth from use, we carefully tapped at the brittle, rust-tinged bottom of the cistern until a jagged hole appeared. We returned the hammer to its peg in the workshop, ensuring it faced precisely as we had found it, with the claws pointing to the right, just the way Dad liked it.

Next, we hauled the pole over to the cistern, maneuvering it through a small opening at the top and threading it through the newly made hole. It was then aligned along the length of the cylindrical structure, ready for our adventurous experiment.

The chair was a challenge. We dragged it over, its legs scraping and catching on the uneven ground, and managed to wriggle it through the opening in the cone-shaped top of the cistern. Perched over the pole, it served as a precarious seat, allowing one of us to climb inside and play the fearless pilot while the other pushed from outside.

John clambered in first, grinning as he settled into the wobbly seat. I stood outside, bracing myself against the cool metal surface, and began to push, rolling the contraption with effort. After some time, sweat dampening my brow, I yearned for a turn in the pilot's seat, eager to swap roles and experience the thrill of the ride myself.

As soon as I climbed into the chair wedged inside the makeshift cockpit, the sharp tang of rust tickled my nose, and the cold, jagged metal dug into my thighs. I gripped the pole with both hands and tried to imagine myself at the controls of a gleaming spacecraft, ready for liftoff. But before I could even start my countdown, the cistern stubbornly refused to budge. I glanced over my shoulder, expecting the jolt of acceleration, only to find that John, determined though he was, could barely nudge the massive cylinder an inch at a time. He shoved with all his might, tongue poking out in concentration, but the only result was a faint squeal as the cistern grudgingly edged forward, threatening to tip before settling right back in place.

"C'mon, push harder!" I shouted; the echo strangely amplified inside the thin shell.

"I am!" John protested, sulking a little. He was younger, but not by much, and the idea of being outmuscled by his older brother stung his pride. Sweat beaded along his brow, and he tried again, this time using his bony shoulder as a battering ram against the dented metal.

The cistern finally lurched forward half a foot, nearly dumping me headfirst into the rusty bottom of the cistern. I righted myself and realized this arrangement was hopeless. We needed more power. I reluctantly extracted myself from the "cockpit," scraping my knee on an exposed screw, and crawled back outside to assess the situation. John was panting, hands on his hips, scowling at the stubborn contraption as if a glare could make it lighter.

We circled the cistern, studying it from every angle. It sat there, a hulking, immovable beast, so much heavier than it had seemed during our earlier games. Even if we both pushed together, the thing would never reach maximum velocity, and my dream of breaking the sound barrier in our backyard would stay just that: a dream.

"We need more speed," I said, feeling strangely like a scientist in a lab. "Otherwise, it's not even fun."

John nodded, but I could tell his confidence was shaken. "Maybe if we got Mom to push?" he suggested, half-serious.

"No way," I said immediately. "She'd just tell us to stop before we even started."

We brooded in silence. Overhead, a vulture circled lazily, waiting for something to die. In the distance, a pickup truck rattled down the road, scattering dust clouds. The smell of cut grass drifted on the breeze, and for a moment, the world seemed content to watch us puzzle out our dilemma.

That's when I noticed the incline at the edge of the yard, the one that sloped gently toward the creek bed below. It wasn't steep, but it was enough. If we could leverage the cistern onto that slope, gravity itself could be our engine.

I grabbed John by the arm and pointed. "If we get it over there," I said, "we can just let it roll. All you have to do is push."

"Won't it go too fast?" John asked, voice tinged with awe and fear.

"Not if we aim for that big tree," I insisted. "The tree will stop it before it rolls over the bank of the creek."

John chewed his lip for a moment, then nodded, his old enthusiasm returning. Together we wrestled the unwieldy barrel to the top of the gentle slope, pausing every few feet to catch our breath and marvel at the risk I was about to take. We lined it up with the tree, an old pecan with a trunk as thick as a city water main, and I climbed back into the chair, my heart fluttering with wild anticipation.

I'm sure you can guess what unfolded next. John gave me a push, and I started to roll down the hill. At first, the ride was exhilarating, a thrilling rush that made my heart race with excitement. I closed my eyes and let my imagination soar, picturing myself as Commando Cody, Sky Marshall of the Universe, the daring hero from the beloved science fiction serials we kids adored in the 1950s. For a moment, I felt invincible. However, the ride soon began to falter. The once-smooth motion became erratic and unpredictable, and a sense of unease began to creep in. Things started to go awry!

The cistern thundered down the hill, picking up speed with every terrifying second. Panic surged through me as I peered through the jagged, rusty hole, my heart pounding. There was Johnny, sprinting desperately alongside the hurtling capsule, his face a mask of determination and fear, desperately trying to position himself in front of the unstoppable force.

"I'm sorry! I'm sorry! I can't stop it!" he screamed, his voice raw with panic. The whole contraption was now hurtling over the jagged rocks; the termite-ridden pole snapped violently in half. The chair and my fragile, young body plummeted headfirst toward the earth, spiraling wildly, screaming and flailing in utter desperation as the slope became steep.

My makeshift spaceship careened faster and faster. One moment, my head scraped against the harsh sides of the cistern, the next I was flung skyward. I reached desperately for the shattered pole, but my fingers grasped nothing but air.

This couldn't go on much longer. I was certain I'd crash into the looming tree at any moment, crawling out with a bloodied head and a body covered in bruises. But fate had other plans. Instead of a direct hit, the cistern veered off-course, slamming into the tree's bark, sending the runaway death trap spinning wildly, end over end. Mere "microseconds" later, I found myself in a surreal state of weightlessness as the fragmented pole, the battered cistern, and the mangled chair plummeted in a dizzying parabolic arc into the depths of the rocky, dry creek bed. Wham! I felt the impact shudder through me, and a sharp pain exploded in my head before everything went dark.

I dragged myself onto the rocks, tears streaming down my face as I fought through the daze and searing pain. I hoped Dad would wait to whip me until I stopped bleeding.

Later that night, when our father returned and found out what we had done, he punished us with his belt, focusing most of his anger on me. It seemed unfair, perhaps because John was smaller, or maybe because I was the one who had hatched the plan. That marked the end of my imaginary journeys to space.

CHAPTER 23

Capoté Knob

The second prominent point on my triangle, west of Highway 80, was barely noticeable from Belmont. It was an anomaly that rose above the flatness of the sandy plains and post oak trees. If you knew where to look and how to notice it, you could see Capoté Knob towering above the horizon from more than 20 miles away. Geologically, it wasn't much to see. But culturally, in the minds of all my Dewville relatives, Capoté Knob stood out like a miniature hell, a place you'd never willingly go. Blanch Lakey Palmero, my grandmother Mimmy, with the mangle scar on her arm, probably feared the knob and passed down that fear and excitement to me more than anyone else. The Capoté Hills, how she called the Knob, was a place teeming with vicious Mountain Lions (panthers) and was the preferred hideout for fugitive outlaws and devious, evil spirits.

In the early 20th century, Mountain Lions were plentiful in the rural areas of South Texas. Both my Palmero grandparents told stories of close and terrifying encounters with "Ole Tom (their moniker for cougars)." When Mimmy was a young girl, she and her sister were carrying a bucket of molasses down the dirt road in front of their house when they met with a crouching and slinking Ole Tom. Terrified, the girls dropped the bucket of molasses and ran for the house. They didn't look back until they were safe inside. Mimmy claimed that the only thing that saved them was the sticky, dark brown molasses, which the big cat briefly stopped to investigate.

My grandfather, too, had a hair-raising story of an encounter at night on the sandy road to Seguin. His story of driving on the dark unpaved country road in his old Ford Model-T was a fascinating adventure tale. Simply getting the Model-T ready for night driving was a task. Grandpa tried to explain the system to me when I was little, but I'm sure I didn't fully understand. I'll leave most of the snorts out this time.

"Well, Bobby, we didn't have any electricity in them old cars back then. We had gas lights. The front headlights were carbide lamps. They burned acetylene gas. The first thing I did when it was getting close to dark was start up the carbide generator, which was mounted on the running board. I turned a little valve to start a water drip onto a pile of calcium carbide. That water caused acetylene gas to come up from the carbide. Once the gas was spewing from a small nozzle in each headlamp, I'd go around with some matches and light a flame. Before you started on a night trip, you had to check how much carbide you had. If you was low, you might end up driving in the pitch dark."

"Grampa, did you ever have to drive in the dark like that?"

"Well, no, but one night I lost one of my headlights. That was long ago, when the road to Seguin was a dirt road. I started out for Seguin after the sun was down, but I could see a little. I stopped just past Big Sandies Creek and flared up the headlamps. I was doing good until I reached that big deep gully on the other side of Goat Hill. I was still more than ten miles from Seguin. All of a sudden, that's when it happened."

Grandpa cleared his throat, snorted, and fell silent. He wanted to withhold the remainder of the tale to build excitement and force Johnny and me to ask him to continue.

"But Grandpa, what happened? Did the lights go out?"

"Not both of them, but one did, the one on the right side. Them lights was pretty dim, and I didn't see him. But there he was. It was Ole Tom in the ditch beside the road.

My mental image of Ole Tom was of a lion, like the ones I'd seen roaring at the start of black-and-white movies. "Did Ole Tom have a bunch of long hair around his neck?"

"No, them cougars like around here don't have that hair. Those ones with hair come from Africa."

"But Grandma said the one that got her molasses had that hair."

"Well, boys, that was her imagination. She was just a little girl when that happened. I've tried to tell her different, but she still says Ole Tom had a hairy neck. I just let her go with that."

Grandpa was a gentle person, and it was fitting for him to allow Mimmy to believe she saw an African Lion. He shifted in his chair and continued the story.

"Well, like I said, that big old cougar was hiding in the ditch, and when I came driving by, he reared up on his hind legs and leaped for my old car. I was down in the low part where the running water had ruffled up the dirt road. It happened so fast that I couldn't do nothing, but I was scared, that I was."

He shifted in his chair again, "Ole Tom grabbed my right headlight with his big old claws and ripped it right off the car. I pushed the lever to speed old Mary up, but couldn't see behind me to know if he was coming on after me."

Both Johnny and I were frozen with excitement, "Now, boys, them old Model-Ts was full of levers and pedals, not like cars today. It had three pedals on the floor. The left pedal was the gear selector. You pressed it all the way down to get in low gear. Halfway down was neutral, and you released it to be in high gear. The middle pedal was to put you in reverse gear, and the right pedal was your brake."

"But Grandpa, what was Ole Tom doing?" He was telling us things we didn't understand and not getting to the good part."

"Now slow down, just slow down now. You got to understand what I was dealing with. It was dark, and I was scared, and my heart was beating like a drum. There was two levers on the steering column, one to speed up the engine and the other was the spark advance. There was also a brake lever. Instead of putting Old Mary in high gear, I pushed the brake pedal all the way down. That made me stop right at the bottom of the ditch. I was sure now that I could hear the big cat coming for me. I let up on the brake and pushed the throttle level all the way down to speed up. I was two miles on down the road before I realized that I was driving in low gear! Full speed in low gear was slow enough for Ole Tom to catch up and grab me if he'd wanted to.

Now, boys, I've heard Ole Tom many times since then. Even right out by the wellhouse in the back yard, but that was the only time I'd seen him in person.

As a young teenager, I was somewhat obsessed with daydreams of adventures outdoors. I felt most comfortable being alone in the woods, by myself, and as far away from people as I could get. I rarely considered the possible dangers I might face. Having lived primarily in the flat areas of Texas, I loved the sandhills to the north of my grandparents' small home on Lakey Road. On the few occasions that Mimmy and Grandpa needed to travel the 18 miles to Seguin, we would take the dirt road through those hills that mesmerized me. The highest hill along road 1117 was the one we all called Goat Hill, because the owner of that hill kept a large flock of goats on the property. But Goat Hill paled in my mind when compared to my visions of the Capoté Knob. I longed to go there, climb its steep banks, and look down from the top. One day, I worked up the courage to ask Mimmy if I could hike up to the Hills and see what they were like from up close. She was out in her extensive vegetable garden gathering tomatoes to can them for the winter. I asked her if I could help, but I didn't tell her what was on my mind at first. I waited until her basket was almost full, and I knew she would be going back to the house.

"Mimmy, I want to do something awful bad, 'cause I like to see things up close. I want you to let me do it."

She pulled the last fat red Manalucie tomato from the vine, "So what's being on your mind to do? There's plenty of work around here to keep you busy. You could take the slop bucket out to the pig or maybe help your Grandpa catch some of those pesky gophers that are digging up my garden and eating the tomato roots. Ain't that enough to keep you busy?"

"If I help Grandpa today, can I go over and climb up on the Capoté Hills? I'll be careful and I'll watch out for snakes."

"Lord, have mercy, boy. You know how far them hills is? That's a fool idea, for sure. And what about Ole Tom? You'd make him a good meal."

"I was thinking about that, and if I see him, I can climb a tree fast. But Grandpa said he thinks he mostly sleeps in the daytime."

"Lord, boy. Don't you know that Ole Tom can climb up a tree faster than a squirrel? He'd get you out on a limb, reach out with his arm, and sink his claws right into your skinny legs. He'd drag your neck right up to his mouth and wrap his lips around your neck. Then, you'd feel his fangs squeezing down with the blood running. It won't take long, so you ain't suffering much. Your hands and legs is going limp. He'll climb back down that tree and put you on the ground. He'll eat all he can, mostly starting with what's in your belly. But if he gets full, and some meat's left, he'll drag brush limbs over and cover you up so he can finish you off when he gets hungry again, next day."

"Grandpa says that if I see Ole Tom, I better not run. I should hold myself up tall, wave my arms, make myself look big, and start yelling at him. If there are some rocks around, I should throw them at him."

"So, you already talked to your grandpa 'bout this? And that's what that old man said?"

"He said he didn't need me to help him with the gopher traps, and he said it would be alright with him, but I had to ask you first."

"So, he put this all on me, did he?"

"But Mimmy, you know, I go around here by myself, don't have any trouble. I've walked over to Uncle Willy's, and I go up to see Uncle Mancel and Aunt Bettie, too. And that's even more walking. I wouldn't get lost. I can see the hills as I go."

"Well, I suppose it will be okay. Boys have got to learn the woods. But you've got to take a sandwich and plenty of water and a good stick, in case you run up on a snake. And you can't stay up there too long. I want you home before it's dark."

My heart was beating with excitement. Mimmy and I walked back to the house. She said one more thing.

"You have to leave early, and you have to stop at Bettie's and Mancel's to tell them what you're doing. And when you get to Aunt Hester's place, you tell her, too."

CHAPTER 24

Unfortunate Bluetick

I was so excited about Mimmy saying it was okay for me to go over and climb up Capoté Knob that I had a terrible time sleeping. When I got drowsy, I would hear some sound either in or out of the house that jerked me back wide awake. Mimmy and Grandpa had an old-timey alarm clock in their bedroom that ticked loudly every time the mainspring released some energy. That alarm clock was not the only clock in the house. In the living room, Grandma had a big wall clock that ticked, too. At the beginning of every hour, the thing would strike a bell as many times as the hour it was. Their house would creak and crack as the loose boards cooled down from the day's blazing heat. I could hear Grandpa snoring with the rhythm of someone cutting a pine board with a handsaw.

In the summer, we all slept with the windows open so that the cooler night air could circulate through the rooms. That night, the wind blew from the west, carrying the sour odor of kitchen waste mixed with pig feces. Occasionally, the sound of a distant cow prompted me to recall a song that Grandma sometimes sang. The only words I could remember were, "The cattle are lowing, the baby awakes, but little Lord Jesus, no crying He makes." Then a pack of coyotes would howl. These distractions, with my barely controllable excitement, conspired to keep me from dozing off. I did fall asleep a few times, but never a deep, restful sleep.

There was another sound from outside that kept me awake. It was firecrackers popping all night long. Grandpa had rented out their 20-acre pasture to a local watermelon farmer, and the deer were jumping over the fence and were biting on the almost ripe melons. The buyer wouldn't buy the melons with tooth marks on them, because the grocery store couldn't sell them. To keep the deer away, the farmer would string long sections of firecrackers on top of the fence. He put a match to one end of the string when it got dark. The first explosion would pop. Then, the fuse would burn for a couple of minutes and reach the next firecracker. It would pop and start up the fuse to the next one. This went on all night and mostly kept the deer out, at least for a while. Grandpa said it was a stupid idea, because the deer learned it was not guns making those pops. After about a week of hearing the firecrackers, the deer used the sounds as a dinner bell.

From my last fitful patch of sleep, I awoke while it was still dark outside. The radio in the kitchen was on at a low volume, and I knew Grandpa was up and listening to the news. I could smell bacon frying and Mimmy getting plates from the cupboard. They were 1800's people, and I never knew them to linger in bed beyond the dawn chorus provided by the summer birds. I got dressed and splashed cold water on my face. I did not have boots in those days, so I put on my ragged old "tennis shoes," that we might call sneakers today. I made my way to the kitchen. Mimmy was at the stove, and I was delighted to see that pancakes were on the breakfast plan. Bacon and pancakes would keep me going all day, but I knew she would be packing me a sandwich and some Fritos, anyway.

"Well, Lord boy," she turned to look me over, "I see you didn't change your mind about climbing those hills. If you're still going to go up there, you better start out as soon as you eat these pancakes. You know it'll be hot as blazes before 10 O'clock."

I said I would and that I'd be careful, and I'd take a jug of water, and I'd eat some bacon if I got hungry. Grandpa turned down the radio a little and gave me his advice.

"If something happens and you get too hot, you go to the nearest place you can find. Remember, those German boys who used to ride the school bus with you live near the Hills. It may not look so, but there's folks living on all the ranches around. You can find somebody if you need to."

I repeated the same thing I'd answered to Mimmy. I'd be careful and safe. I wolfed down my breakfast and packed some biscuits, a bacon and egg sandwich, and two Mason jars of water into a tow sack and started on my adventure.

It took me longer than I thought it might to make it to Aunt Hester's place on Big Sandy Creek. The creek was always dry, except during periods of heavy rain upstream. I was a little disturbed that I'd lost sight of the Hills. The creek was in a valley, and the trees were tall and close together, so I tried to keep walking in the same direction until I could see the Hills again.

I walked for a while before finally spotting my destination in the distance. It was getting hotter, but I wanted to save my water for as long as possible. I had crossed over three or four barbed-wire fences and came into an old corn field. Most of the brown stalks were bent over and touching the sand, which made walking difficult. The good thing was that I was getting closer to the Hills, but they didn't seem as lofty as I thought they should have been.

Up ahead, I could see some buzzards (that's what we called vultures) and a Mexican Eagle (caracara), fighting over the putrid meat of some unfortunate animal's body. The birds got nervous when they saw me. They were reluctant to lift off the ground but watched me closely. As I got nearer, the vultures finally took to the air, long wings pulling their gorged bodies toward the nearest trees. The feisty caracaras waited until I could have hit them with a rock. They reluctantly headed for the trees and joined the vultures. I was sad to see that the remaining carcass was part of an unfortunate bluetick coon hound that some hunter would never see again.

I stopped for a minute and sat on a big rusty-colored rock to drink some water. The area was becoming increasingly dense with trees, brush, and weeds. There were what we called cedar trees, as well as many shrubs with orange-red berries. Like the one I was sitting on, many red rocks protruded from the sandy ground, some small and some large. The sky was filling with clouds, some looked like they were building up to be showers. At least it was cooler under the cloud cover. I wouldn't have minded if it rained a little, but I was afraid of lightning. It must have been raining somewhere, because I could smell the dampness on the increasingly gusty breeze. The birds had stopped singing.

I didn't have a watch, and with the cloud cover, I could not see the sun. I was pretty sure it was sometime after noon, and I felt a sense of disorientation. Okay, I was lost. The hike was turning out to be more difficult than I had anticipated. I began to worry that I wouldn't be able to make it back home before dark. I was a stubborn kid and prone to taking risks, and this idea of hiking to the Hills was building into yet another grave mistake. I put my water away and started, with more urgency in my steps, in the direction I assumed was the shortest way to the safety of Mimmy and Grandpa's place. I searched around for an opening in the brush large enough to get a glimpse of the overall landscape, but the vegetation continued to close in on me. I usually felt very comfortable and safe in the woods, but this time I was fighting back a sense of panic. I started scanning every tree for Old Tom, just waiting for me to pass under his perch. I was lost and had no idea which way I should go. My boastful confidence meter was now at zero.

Owing to my inability to see more than a few yards ahead, my brain magnified the few sounds I heard. I thought about a television show I had seen where a lost person determined his direction by noticing on which side of a tree the moss was growing. Moss did not grow on the sides of Post Oak or Blackjack trees, or maybe any tree in South Texas. I thought of what Grandpa said. If I get lost, try to find a family's house on one of the nearby ranches. I reasoned that going downhill, rather than uphill, would be my best chance of finding a home or a road, but due to the dense brush and trees, it wasn't easy to decide which way was up and which was down. I picked a direction and focused hard on continuing in a straight line. I would select a tree ahead, keep it in my vision, and walk to that tree. From there, I'd choose another tree and work my way to that one. Even with this method, I don't think I was able to travel in a straight line successfully, but the process did help settle my sense of panic.

I heard a distant sound and began talking to myself out loud. "I think that is a car or truck driving on some road over that way." The sound came from a different direction than the one I had randomly chosen. But the sound of a vehicle, I think a truck, meant there must be a road or a house to my left. Maybe it was a tractor, and some farmer was plowing his field. Yes, it must be a tractor because the sound was coming from the same general direction, and I rushed as fast as I could toward it. If it had been a car or truck on a road, the amplitude of the sound would have diminished after a minute or two. If the farmer continued to plow his field, or whatever he was doing, I could follow the sound in a straight line. I knew I should keep moving as fast as I could.

I kept moving, working my way toward the tractor, weaving through the trees and brush – and then disaster struck! I came to a long clearing through the dense vegetation, a clearing about twenty feet wide. Right down the middle of the clearing was a 10-foot-high deer fence with six feet of hog wire at the bottom and several strands of vicious-looking high-tension barbed wire at the top. I certainly could not climb over this fence, and since the hog wire was buried in the ground, crawling under was not an option either. My sense of panic had become elevated again, but it was now even higher. I pushed myself against the fence, gripping the hog wire in both hands, trying to decide what I should do now. Moving forward was impossible, and going backwards was the worst of my diminishing options.

As I leaned on the fence, the sky slowly darkened, and I felt a few drops of rain falling on my face. The wind was picking up, and the sound of the tractor began to fade. I couldn't tell if the tractor was now moving away or if the wind and light rain caused the diminishing sound.

Fences had to have a gate, but on these ranches, I knew most gates were locked unless someone with a key or combination was passing through. At least, there had to be a road. But which way should I walk? If I could get to a road, someone would eventually come along and find me, even if I had to wait all night. That would be my safest option, but which direction along the fence should I go? Right or left? There may only be one gate in this fence and, if I chose the wrong direction, it would be awful. Some of these ranches were thousands of acres, with a multi-mile-long wire wall to walk around, a daunting prospect in the looming darkness. I had to make a choice. Something inside me spoke and said, "Turn to the right." That's what I did.

Having made my decision, I picked up my pace. All the trees and brush were cleared from the fence, making walking easy, but with urgency. Up ahead, I could see a flock of wild turkeys ambling around, feeding on acorns. The second they noticed, they jogged into the brush and Post Oaks. I was getting very thirsty and decided to drink. As I walked, I finished off the remainder of my water supply. It was easier to carry the hydration in my belly and not in a jug.

I continued along the fence for at least an hour and saw no gates, nor any sign of recent human activity. I'm guessing it was about six in the afternoon, and my grandparents would be concerned about me. They would have been calling their neighbors to see if anyone had seen me. It is hard to believe, but back when I was lost and walking that deer fence, my grandparents' phone looked like something Alexander Bell would have used. It hung on the kitchen wall and had a crank handle for dialing. Everyone around Dewville and beyond was on one giant party line. If Mimmy wanted to call Aunt Betty, she would put the earpiece to her head and crank in Aunt Betty's code. I don't remember her code, but it was something like a short ring, two long rings, followed by a final short ring. If we were using Morse Code, that would be a "P." I assure you that I was not thinking of Morse Code while walking that fence. I would not have been telling you about old telephones. I was scared.

Then my heart began to beat faster. Up ahead, I saw a car driving toward the fence. That meant there was a gate, at least I assumed so. I was still far from that presumed gate and the moving car. I started running as fast as I could. I was yelling and waving my right arm to get the driver's attention. I was close enough now to see the gate. The car stopped, and a boy, maybe a teenager, got out and opened the gate. Another person, maybe his mother, drove through the gate. I ran faster and yelled as loud as I could, but he did not see or hear me. This wasn't good. Before I got in range, the boy had closed the gate and hopped back in the car. They drove away without seeing or hearing me.

I was out of breath and filled with panic. I collapsed on the ground. If I had not stopped to swizzle down my last measure of water, or if I had not spent all that time deciding whether to go right or left when I found the fence, I could have been waiting at that gate. I felt like crying, but I told myself I was too old to cry. Instead, I forced myself to get up and walk to the gate. I sat down beside it and watched the sun set through a thin layer of clouds. I could hear a pack of coyotes howling from a distance. I needed to think. Should I stay at the gate and wait for someone, perhaps the mother and son who left the ranch, to return, or should I start walking down the dirt road away from the gate and hope to find a house or make it to a road or highway? I was sure that Mimmy and Grandpa were frantic and had called everyone they knew, so people would be looking for me.

I decided to stay at the gate. There was a sense of safety at the gate. It was something human, and sooner or later, someone would come through that gate, and I'd be right there waiting. I'd forgotten entirely about Capoté Knob and the great adventure I might have had.

It grew very dark, and the coyotes continued howling. Their yodeling voices started up some howling from another pack in the woods closer to me. I wasn't all that afraid of coyotes, but they prompted me to think of rabid dogs. Coyotes were like dogs, and I knew they could get rabies, too. If one got after me, I could climb up the gate and be safe. At least that's what I told myself.

I don't know how long I waited there, but the boy and the woman had not returned to the ranch. It occurred to me that they may have been visiting the ranch and would not be returning anytime soon. What a stupid idea this adventure had been. All the bad outcomes that my grandmother had warned me about had failed to dissuade me. Now I was alone, exhausted, hungry, frightened, in the dark, and possibly surrounded by rabid coyotes. I would never want to go to those hills ever again; that was sure.

"What! Was that a car I heard coming up behind me to the gate? It was! I jumped up, and I could see the headlights through the brush. An old, beaten-down pickup truck stopped at the gate, and the driver saw me waving and got out.

"¿A dónde vas, chamaco? Súbete, no es bueno andar solo en la oscuridad."

I did not understand a word, so I answered in English, "Mister, I was in the woods and got lost, and I don't know where I am."

He answered me back, "I go Nixon. You want come, too."

I shook my head, and he locked the gate, motioning for me to ride with him in the truck. I think he knew very little English, so we rode along the bumpy sanding track in silence. Finally, we reached Highway 80 and turned south. We passed Mr. Kidd's little store in Leesville, and I knew we weren't far from Nixon. I could get someone there to call out to my Grandparents' house, and Grandpa would come to get me. The farmhand dropped me off at a gas station in Nixon, and I asked if the man there would call for me. I didn't know that it was complicated because of the connection that needed to be made from Nixon to Dewville. The service station man was required to contact an operator in Leesville who had a switchboard in her spare bedroom. She was able to ring up Mimmy with her designated ring code. When my grandmother answered, she was frantic. The operator used a short cable to connect Mimmy to the man at the service station. Grandpa was sent to get me.

I thanked the service station attendant and sat on the pavement beside a gas pump, waiting. In about 20 minutes, Grandpa arrived in his faded green 1949 Ford, a car that looms large in my memory. We rode back home in mostly silence. Grandpa did not scold me or question me. Neither did Mimmy. I think they understood that I'd learned my lesson. They were good people, and I loved them.

CHAPTER 25

Over The Moon

January 18th, 1970. The first day of classes for the Spring term at Southwest Texas State started this day. As had been my life's pattern, I had not slept well. Before daylight, I was awakened by the patter of light rain against my bedroom window. It was dark and cold outside. I could feel the warmth of Carol, my girlfriend, lying next to me. She was sound asleep, warm and breathing quietly. Her blonde hair fell lightly on my shoulder. I didn't want to get up.

I was the only one awake in the old house on North Street. It would be hours before Deco would stir. I lay pondering my polarized life and its future. By "polarized," I mean my lofty aspirations at one end and my shadowy monsters, omnipresent at the other. I was enrolled in General Chemistry, Analytic Geometry, Mechanics,

Electricity, and the History of the United States. This was a heavy load, but exciting. I felt ready to pour my soul into the work.

Behind me in the dimness, but looming, stood my failed marriage and its obligations. Rachel's divorce proceedings had been approved and signed by a judge over the Christmas break. Along with that came a court order to pay child support, an amount that I would struggle to earn each month. The saddest part was my limited visitation rights with my young daughter, Tracy. Clearly, I was not good at this marriage thing.

Compounding my perils was Carol Kidwell, breathing so softly beside me. She was four months pregnant. A baby girl was to join us in July. Carol worked as a waitress and contributed to our income, but I understood that this could not continue during the last weeks before Brandy Ann arrived. I tried to shutter most of this into a closed chamber of my mind, but being hidden did not solve the problems. With the baby, I'd have to figure out how to afford an apartment. Living in Deco's old house was not an option. These elements were at the gloomy negative pole, a direction I forced myself not to turn around and examine. At least not on the first day of classes.

I crawled from the warm bed and made myself some coffee. The house was quiet. I was careful to make as little noise as possible. I forced down a couple of slices of bread, even though I didn't feel like having breakfast. I put on my tattered jacket, still splattered with remnants of brick mortar, and started my uphill walk to campus. The short collar didn't protect me from the cold droplets landing on my neck.

It was only a block or two, uphill, to a point where I could see Old Main. Its appearance brightened my malaise. I really was looking forward to my first day of classes. Like last semester, I would have a class with my brother John, a physics class on Electricity and Electronics. I wouldn't be worried about reading a passage about large penises. The embarrassment in speech class was still raw. For those of you who haven't lived such a contrast, you can't imagine how exhilarating being in college was. And the physics department was the best possible world for me. It was a small department; five faculty and about 16 physics majors. It was certainly a different time and a different way of teaching students. Physics majors were given free run of the physics department's space in the science building. I don't remember if we were issued keys, but I know it was common for us to be buzzing around the classroom and lab at all hours of the night. That's where most of us did our homework. Physics homework was always a set of problems to solve. There were no reading assignments or writing assignments; only physics problems to solve. We would swarm in and stake out a room with wall-to-wall slate blackboards, and a group of three or four of us would work together. Other groups would be settled into other classrooms, tackling the same assignments. We might be at this until two or three in the morning. If the assignment contained a particularly difficult problem, you might hear yelling from down the hall, "Hey, guys. We got it. WE GOT IT!" We'd all rush down the hall to see a solution on the board. What camaraderie. What freedom! What elation! This was the best time a human could experience. It felt like a religious conversion; a come-to-Jesus moment (something I've never had).

Since our department was so small, physics undergraduates were recruited to serve as teaching assistants for the large general physics classes that biology students and others were required to take. We had an office with several desks on the physics floor. This arrangement was like no other discipline I know of at SWT. A different time and a different world, those days were.

I remember one amazing day during the spring of 1970. As I sat at my desk working on homework, another teaching assistant entered with a friend named Robin Whitaker and introduced him to me. Robin had graduated from our department the previous Spring and was back for a visit. He was a little overweight and had a thin beard. He seemed friendly.

He reached out to shake my hand, "Nice to meet you, Bob. That was my desk last year. I hope I left it clean enough."

I smiled, "Haven't found the solution to this problem stuffed in a drawer. Did I hear you say you are at Rice now?"

"Yep, I'm in the Space Science Department working on my Master's. Working on Apollo stuff."

"Wow, I'd love to be doing that."

"Hey, it's a lot of fun. I'm glad I got to go to Rice. We're close to NASA, so they're tight with the Apollo Program."

We chatted for a bit longer, then Robin walked down the hall looking for other friends to visit. My office mate was grinning from ear to ear.

"What? What's got you so happy?"

He looked around with an exaggerated neck turn in both directions.

"No. It's nothing. I'm just glad to see Robin again. As he said, we used to be in this room together."

He kept reaching into his pocket and fidgeting. Something wasn't normal here. He sat at his desk and continued to feel around in his pants pocket. He looked like he was bursting to tell me something.

I stood up and walked around his desk. "Okay, Jerry. What's in your pocket that you keep fondling down there?"

I want to show you, but you have to promise that you will never tell anybody what you saw. Never."

"Hey, you know I'm not going a say anything, so pull it out."

He pulled the object from his pocket but kept it hidden in his curled fist. "You absolutely promise?"

"Come on now. You're acting like some stupid kid. I already know you've got something, so just open that fist and let me see, damn it."

Jerry slowly uncurled his fingers. I looked. All I saw was a tiny grayish speck. I bent down to get closer, but it still looked like something he'd dug from the heel of his shoe.

"Okay. I see a grain of something. What is it?"

Jerry lowered his voice, "It's a piece of the moon."

"What?

"It's a piece of the moon. It is. Robin said he was working with some moon rocks collected during the Apollo 11 flight. He said, somehow, a small chunk, just a little piece, dropped into his pocket from the workbench. I know he was lying about how he got it, but we've been friends for a long time. He wanted me to have a tiny piece."

I wanted to touch it, and Jerry let me. I had trouble believing it was true. Had I really touched and held a grain of regolith from the moon? I dropped the fragment of gray stone back in his hand. I felt truly unique.

"Look, Bob, you know you can't talk about this. I don't know if it is legal, so I don't want to take a chance.

I watched Jerry place his moon fragment between two microscope slides and apply tape around the edges so it wouldn't slip out. He put it in the top middle drawer of his desk. Maybe someone knew it was in his desk, because it disappeared within a couple of weeks. Maybe Jerry did something with it and just told me it was missing. All I know is that I had it in my bare hand for a while, and it left me with a profound sense that I had experienced something unique. I had touched, with my own skin, a piece of the moon.

I cannot express what a profound, deep inner feeling, touching that fragment of moon rock created. All my senses sizzled alive, like I had stood and observed the Universe from a God-like perspective. When I was a child, I had broken flint rocks with a hammer and examined their insides. I knew that I was the first and only human ever to see that inside surface. Just me, and me alone. No other human had seen it before. The moon rock was like that.

Just so you know, on January 20th, 1970, Carol Kidwell and I went to the Hays County Courthouse and got married. Twenty-two days after my marriage to Rachel Smith ended. Physics I could learn. Being good at marriage, I was failing to learn.

CHAPTER 26

Green Gas

The freedom that physics students had at SWT in the 1970s was breathtaking and euphoric. We lived a life that would be unimaginable today. But it was also incredibly dangerous. I've lived a "charmed" life. I've flirted with death so many times and survived events against all odds – way more times than any nine-lived cat could boast. This survival marvel is another unspoken mental argument I apply as justification for my skepticism about religion. Would a vengeful God allow a non-believer and a hard sinner so many escapes from death?

My brother and I were snooping around in the storage room of one of the physics labs one night and discovered a super-neat 1930ish refracting telescope. I don't remember the diameter of the objective lens, but it was at least six inches. It had probably been unused in the storeroom for decades. As if we didn't have other crucial things to do, like studying and homework problems, we decided to restore this relic to its former glory. What really looked ugly were the several brass parts on the instrument. They were corroded and pitted. We found some sandpaper and tried to brighten up the brass, but that wasn't working well. This sanding was hard, slow work, and I didn't have the patience for it.

I looked up, but continued sanding a hefty brass ring, "Hey John, this is bull shit. We could be getting this one piece cleaned up all night. There has to be a better way."

"Here, let me take over for a while."

I stood up from my stool and let John take over the sanding. "It would be super cool if we could get this thing looking pretty and working again. We could show it to Dr. Anderson when we get it all polished and looking new. We may have to find some white touch-up paint, too."

John nodded his approval. Always curious, I started snooping through the cabinets under the lab's various sinks. I checked the clock in the room, and it was after midnight. I think John and I were the only humans in the building. I finally got to the big sink, with all manners of ugly stains, at the back of the room.

When I pulled open the doors under the sink, there were several large, heavy Pyrex jugs of various acids. There was about a gallon of sulfuric acid, also labeled Oil of Vitriol, and a jug of nitric acid, labeled Aqua Fortis. There was about half a jug of hydrochloric acid, labeled muriatic acid. I had had very little experience with strong acids, but I remember Grandpa using them around his blacksmith shop at the state hospital. He was unafraid of these liquids and was casual about their use. The labels on these acid bottles under the sink weren't all that scary looking. They just said Corrosive.

Thinking about Grandpa's unconcerned use and the rather tame labeling on these jugs gave me an idea. I called John over. He looked tired of sanding anyway.

"Hey, come look what I've found. This might clean those parts up fast."

John looked under the sinks with caution. "Those are acids under there. We'd need some rubber gloves and gas masks to play around with those. This is a bad idea."

"I don't think anything crazy would happen. Remember, Grandpa used to use this stuff all the time. He doused steel in acid, I think. He called it pickling. It cleaned slag off the steel."

John looked disturbed. "I'm not sure about this, but we could try a little bit first and see what happens."

"Now you're talking. Go get that brass ring you were working on and let's try."

While John retrieved the brass part, I found a rubber stopper and plugged the sink drain. I didn't want to waste acid by letting it run down the drain as I poured it onto the ring. I placed the ring in the sink, then poured full-strength hydrochloric acid over the brass. Nothing very interesting happened. There were a few little bubbles drifting up, but no real cleaning action that I could see. John was standing back and couldn't see inside the sink. I waited for a couple of minutes, hoping to get things going. I found a pair of tongs and sloshed the part around in the acid. Still, no real action.

John asked me what was going on.

"It's not doing much cleaning. I'm going to add one of those other acids and see if they work better."

John nodded his approval but stayed behind me. I selected nitric acid and poured a healthy gulp onto the brass ring, while a pool of hydrochloric acid still floated in the sink. Wow. I was frozen for a few seconds. Things were happening now, that was sure. At first, there was a fizzing sound, and then an ugly yellow-green cloud began to rise from the sink. My nose detected a sharp odor, like a public swimming pool, but much stronger. Within seconds, the heavy green cloud reached the top of the sink and rolled over the edge onto the lab floor. The legs of the desks and chairs were invisible beneath the eerie greenish vapor. My instinct was to run, but I turned to see what John was doing. He was backing up, and serious distress was painted on his face. We both ran for the hallway and closed the lab door behind us.

Looking back through the glass in the door, I could see that the green cloud now completely covered the lab floor. And frighteningly, it was seeping from under the door into the hallway.

"Damn, John, it looks like I screwed this up. We need to do something."

"You're right about that. This may get us kicked out of school. That cloud is getting deeper by the second. Hell, that might just kill somebody."

I knew he was right, and this was all my idea. He'd warned me. I didn't listen. It fell to me to do something, but what?

I was not thinking clearly, but John was.

"Okay, we have to get those windows in there open."

The room was about 50 feet long, and there was a row of windows to the outside where fresh air could be had.

"You're right, I'm going back in there and open those windows."

I took a long, deep breath, held it, jerked open the lab door, ran through the now almost knee-high cloud, and started pushing up the windows. I got about half of them open before I needed a new breath. I ran back to the hall for air. Opening the lab door allowed more of the heavy, billowing gas into the hallway. But it was down around our feet, and we could breathe clean air above.

After inhaling another deep breath, I plunged back into the lab and got the remaining closed windows open. To our great relief, there was a nice breeze blowing outside, and the Yellow-green demon had slowed its ascent from the lab sink. The gas density in the room was decreasing very slowly. In a couple of hours, we felt safe enough to go back into the lab and clean up the evidence of our disaster, or should I say, my disaster. We never told anyone what happened.

We returned to school the next day, and the faculty and others I did not recognize were walking through the halls, trying to find the source of the faint chlorine gas odor. I don't think they ever pinned it down to a room on the first floor of the science building, but they suspected it came from there. No one ever questioned John and me about it. We were lucky twice over. We were still in school and still alive.

I should mention that, owing to my personal problems (money and women) and my lack of preparation for college, my grades in the Spring semester were as dismal as I had anticipated. I withdrew from General Chemistry and the History of the United States. I scored a B in Mechanics and Heat, an A in Electronics, the course I took with my brother John. I failed Analytic Geometry. I was still barely above water, but not by much. I'd need to retake one course, but I'd completed one year of college, almost. I took that as small encouragement. So far, he was wrong, Mr. Staples. I was still in college.

CHAPTER 27

The Snap

My girlfriend Carol Kidwell and I were married. It didn't take long for me to realize that I was not cut out for marriage. The Rachel marriage was a failure, and now this new marriage was producing friction, too.

Over the several days we were moving, piecemeal, to the apartment, we were back and forth. There were several occasions when Carol and Deco were alone while she was packing. At first, I was so self-involved that I didn't notice the friendship forming between them. She needed someone to pay attention to her plight, and that should have been me. This developing familiarity between my wife and Deco was a direct result of neglect. My neglect.

I was slow to notice this growing familiarity, bordering on flirting. It may have been innocent, but stress and something far more toxic and shocking were at play in my twisted mind. Maybe, for the first time ever, I saw my father in me. I came in from Analytic Geometry class one afternoon after failing a mid-term exam, and there were Deco and Carol sitting together on the ragged old sofa. Sitting too close together, as I saw it. A fiber in my brain meant to civilize me, snapped. Within seconds, I felt uncontrolled rage.

"What the hell is this?" I screamed.

Deco, inhaled from his hand-rolled smoke, and squinted at me. He didn't say a word.

The room went dark except for Carol and Deco sitting there together. I went blind. Oblivious. All I could see was what I took to be betrayal. I don't remember the next few seconds, but I found myself on top of Deco, pounding his face with hard, repeated blows. His "smoke" had fallen to his chest and was burning a hole through his shirt. I could smell the cotton fabric burning. Blood was pouring from his bottom lip. Carol tried to pull me away but couldn't. Deco never tried to fight back. Finally, the blood covering his face and neck repulsed me, and I stopped.

I grabbed a packed box of belongings, dragged Carol, pregnant as she was, and left the old house and Deco for the last time. We had a few items left there, but I asked my brother to fetch them and bring them to us. He never said so, but I am sure he was crushingly ashamed of me. It wasn't me beating up on Deco. It was James Albert Benson swinging the punches. I had become my father. Or maybe it was always my father suppressed inside me, coming to the surface.

I never spoke to Deco again. I saw him on campus a few days later from a distance. He looked terrible. His right eye was blackened and partly closed, and his bottom lip was bandaged, likely to cover several stitches. I felt ashamed of what I'd done.

That I could snap like that, become a complete Jekyll and Hyde, was a burning knot in my psyche. How could this monster be buried so deeply in me that I didn't know it existed? But I knew it then, and it crushed my soul.

A few days later, I went to a pawnshop and sold our television and a lawnmower. With my recent GI Bill check and the cash from the pawn shop, we rented a Ryder truck and a towbar. I hooked up our rundown VW to the towbar and moved to Houston. I didn't officially drop my classes. I was spiraling into a cesspool of despair. Jobs and College didn't matter anymore, and worst of all, I was dragging Carol alone with me. As we drove past the San Marcos City Limits, Mr. Staples' voice got into my head.

"Science, huh?" he said, almost chuckling. "You want something that will get you booted out of here quickly, huh? Let me suggest a field of science for you. You should major in physics. That will be a true test for you. If you make it through that, and believe me, you won't, no one will ever question whether you went to high school. That's your perfect major. It will see your butt off this campus by Christmas."

Our child, Brandy Ann, was born on July 4th, 1970, exactly 194 years, to the day, since our country declared independence from British rule. She was a firecracker baby. In 1970, births in the United States totaled 3,731,386. On average, 10,220 babies per day! So maybe Brandy's birthday wasn't as unique as we thought. It takes a lot of childbirths to make ten thousand babies.

Our new baby, Brandy, was a sweet child and seemed happier than most newborns I'd been around. But I was not doing my part as a father. My proclivity to take on too many tasks at once is a deep flaw in me, one I live with to this day. This wasn't fair to Carol or to Brandy. They needed attention that I was unable to provide. I should just own it. I was selfish. I was driven by my own personal demons and pushed aside my domestic responsibilities. I was on edge and filled with a hidden anger that strained my external interactions with the people around me. Was I ever going to succeed at anything? It did not seem likely.

CHAPTER 28

Chester LeBlue

Now here we were in Houston: Me, Carol, Brandy, and all we owned in the back of a Ryder truck. Clearly, my mind had snapped. This complete abandonment of San Marcos, of college, and everything I'd hope to achieve was insanity. The drive from SWT to Houston had been depressing. Carol, sobbing, and Brandy crying were voices I could barely tolerate. What the hell was I going to do now? I had a friend there, and he took us in for a few days until I could find a job. We slept on the sofa and on the floor and batted away his roaches. They were everywhere. They crawled along the ceiling and baseboards. They crept out of the drawer where the silverware was kept. They strode across your face when you tried to sleep. That was Houston; roaches, hot, smelly, crowded, flat, busy, prosperous, and vibrant. I can't remember why I decided on Houston as the escape route.

The next morning, on Thursday, I set off to look for the nearest brick job. Bricklaying is the sort of employment that does not require an agency to screen you or for you to fill out an application. You just put on your hard hat, pick up your toolbox, walk onto the jobsite, and ask someone to point out the foreman. If there was mud on your trowel, you were hired. I got my first paycheck on Friday. My pay was $3.50 per hour. I'd earned $49 minus what the government took out for taxes. That would buy a little food and a package of Pampers. My friend didn't have a way to wash cloth diapers.

My next paycheck was enough to get us off the roach-ridden sofa and floor, and into a small apartment. An apartment with privacy and with fewer roaches. I kept at the bricklaying but longed for something better.

As a child, thinking about birds and radios provided an escape from Dad's unpredictable and sometimes violent behavior. On my way home from work, I'd drive by a radio shop, Metro Radio & Speedometer. I passed the place every day but had no idea precisely what they did. One day, I decided to walk in and look around. That was the day I became a radio technician.

Chester LeBlue, the owner, tried to look respectable, but the act didn't hold up. He leaned on his counter with a toothpick in his mouth and a grin that came too easily. The shop smelled like solder and grease, radios gutted on the bench, and stacks of odometer dials shoved in the corner. He talked fast, offering to fix the static on my radio, then dropped his voice to mention other kinds of work. Folks who came in for honest repairs left uneasy, and the rest knew precisely what he was selling. I must have looked like an easy mark.

"Hey, how ya doin'? C'mon in, don't be shy."

To hear someone speak with that heavy accent threw me off. I guessed he was from New York. Honestly, I had trouble understanding what he was saying. Before this unlikely character registered in my brain, he continued.

"Whatcha got, radio's buzzin' out on ya? We'll take a look, won't cost ya nothin' ta talk about it. Name's LeBlue, you bring it, I fix it. We'll getcha squared away."

I told him I was just looking.

He squinted like he didn't quite believe it. "Just lookin', huh? Kid your age, standin' in a radio shop, sayin' he's just lookin'? Nah, I don't buy that. You like radios, yeah? You like how the transistors feel warm when the juice runs through 'em?"

I nodded, not sure if he was making fun of me.

"Course ya do," he said, tapping his toothpick. "Lotta fellas come in here and don't know a resistor from a rowboat, but you, I can tell you've been peekin' inside the box. You ever open one up? See the guts? Smell that burnt-dust smell when a cap blows?"

I admitted I'd fiddled with a radio or two, nothing serious, just curiosity.

"That's how it starts," he said, lowering his voice. "Curiosity. That's worth more than half the junk sittin' on my shelves. Curiosity keeps a man up late, keeps his hands movin', keeps the work comin'. You got that."

I shifted my weight, suddenly wishing I'd kept driving past his shop.

He grinned wider. "Tell ya what, kid. You come in here a couple afternoons, sweep up the solder clippings, maybe hold a meter steady while my guy checks a coil, see if you like the smell of the place. I'll toss ya a couple bucks. Who knows, maybe you'll learn somethin' that sticks. Whaddya say?"

I hesitated, staring at the mess of parts behind him, the open radios with their wires spilling out like guts. He caught my silence and leaned closer.

"Don't think too hard, eh? Work's work. Better than bustin' your back out in the sun. You come in, you learn, you make a little scratch. Door's open."

I took Mr. LeBlue's offer. The next morning, I stopped by the construction site where I worked and told the boss I'd taken another job. I was going to learn to be a radio technician. He smiled, chuckled really, and said, "Good luck. I hope you do a better job on radios than what you did with those crooked bricks you put into my wall yesterday."

I walked away thinking my bricklaying career was over, forever. That would not be so.

My first day at Metro Radio & Speedometer wasn't what I'd hoped. Instead of learning to repair radios, I rode with another technician who made service calls to car dealerships all over Houston. His workday was fast-paced and simple. We'd pull into a dealership, stop at the service desk, and be handed a stack of work orders for cars with radio problems, usually four or five cars and sometimes a pickup. The guy training me was named Jim.

"Now, let me tell you how this job is done. We only try to fix the ones with simple problems, things like a blown fuse or a loose antenna wire causing static. The boss says, 'If you can't find the problem and fix it in five minutes, you don't fuck with it anymore.'"

All the cars said to have "radio problems" were lined up in the back lot. These cars were all still under factory warranty. Jim examined the first work order. It was a Chevrolet Camaro, in for an oil change and a loose rearview mirror. There was a separate note: radio static when the engine is running. Jim started the engine and turned on the radio.

"Now listen. Do you hear any static?"

I shook my head, "No."

"Well, there ain't none. You're going to get a lot of cars like this one. The ticket says something's wrong with the radio, but when you turn it on, it's not doing that something. After the car owner leaves the car and goes home, the service manager sketched in that extra note - that note about the radio static. He's just cheating General Motors."

He handed me the service order. "So, what do we do then? Do we jiggle the wires under the dash to see if we can hear any static? Maybe some wires are loose, or something."

"Naw, you ain't getting it. We don't have time for none of that. We got seven dealerships to be at before we go back to the shop, maybe thirty radios, all total."

I looked puzzled.

"So, here's the deal. If the ticket says something's wrong, and we see it working perfectly, then we don't do anything. We write down that we tightened the wires, including the antenna cable under the dash. Tomorrow, when you start driving the other van, making your calls, you can crawl under the dash and jiggle the wires if you want, but it ain't hardly any use. You write on the ticket that you checked it, and it's now working. You pull the yellow page out from this form, give it to LeBlue, and he bills the dealer for a radio repair."

"That doesn't sound legal."

"Hell no, it ain't legal at all. We just keep our mouths shut, and we get to keep our jobs. You see, LeBlue and that service manager has got a little private deal going. LeBlue gets paid for these phony work orders. Once a month he comes visiting around, pockets bulging with twenties, to chat up his service managers , splits the cash with them. Hell, that guy Tommy, the service manager here, who handed us these work orders, bought himself a big boat and motor last year with a wad of twenty-dollar bills."

"So, what do we do if there really is something wrong?"

"Look there in the back. See all those radios stacked on those shelves? All of them are working radios pulled from cars. Some of those, the technicians fixed and are now working, and some of them, LeBlue picked up from wrecked cars at the junkyard. So, if there is a real problem, we get under the dash, remove the radio, and replace it with a working one. We don't fix anything out here. All we do is pull the radio and slip in a working one. The technicians back at the shop do the fixing, and they go back in our vans as spares."

This was a disappointing revelation. I'd quit my job that paid more than this one to learn a trade. All I was going to "learn" was how to unscrew a radio, drop it from behind the dash, and install one from a junk yard.

Remember the name of LeBlue's company - Metro Radio & Speedometer? LeBlue also had a thriving business turning back odometers in used cars. I mean, some used car dealers would take a piece of junk car as a trade-in, maybe with 150,000 miles on it. He'd call LeBlue, and LeBlue would send some ex-safe cracker with special homemade tools to turn the odometer backwards to 55,000 miles, or something like that. Some poor sucker would buy the junker, thinking what a low-mileage deal he gotten, and drive away with a big smile on his face.

I didn't work for LeBlue for more than a couple of months. I could have stayed on, but the way LeBlue operated made me uncomfortable. His methods were unethical at best, illegal at worst. Maybe I didn't fit in well with a Yankee boss. Or maybe any boss.

I'm not saying I had squeaky-clean morals in those days. I could have swallowed hard and kept the job, but I was making less money and not truly learning a new trade. At least I could boast that I once worked at a radio shop. I went back to bricklaying.

After a couple of months of stepping over brick bats and climbing up scaffolding, my work boots were wearing out. I needed better boots. There was a Goodwill store in one of the sleazier parts of Houston, where I might find a used pair. It was next to a rundown old building that sold high-premium, high-deductible car insurance, which we would today define as a call center. An office full of low-paid automatons sat at desks for eight hours a day, randomly calling numbers from a telephone book, and using unfair scare tactics to push old people into high-dollar, low-value policies.

There was no parking space left at the Goodwill store, so I parked in front of a little consignment resale shop next door. I noticed a man leaving the call center and loading an odd electronic device into his panel truck. I walked over to see what he had.

The guy told me he'd just installed a new switchboard at the call center and was taking the old model back to his company. There was a technician who would refurbish it, and they'd sell it to someone who couldn't afford a new one.

"These electromechanical piles of shit need replacing often. We have these things in offices all over Houston. Hardly a day goes by without some glitch that requires replacing a relay or something else. I stay busy as hell."

I was curious. "I've never seen anything like that."

"It's just a big pile of relays that connect 50 phones to the incoming phone line. When a customer calls, a recorded voice asks them to dial a three-digit number, and this connects them with a particular person at the company. It's just like the old switchboard, where a woman would answer a call and then connect the caller to the person by plugging in a wire. It's just automatic and does not need an operator."

"It looks complicated."

"It takes a while to learn it, but I've been at it for a long time. It's not so bad, once you see how it works."

I pondered how different a technical job like his was from my job. You didn't have to be smart to be a bricklayer. In fact, you're better off being a little dumb, where your mind doesn't wander off and daydream about other things. Monotony and smarts were enemies. I remembered my father saying once that people with IQs around 80 make the best truck drivers. They can focus on the road without being distracted by a wandering mind.

I bought my new used boots and thought about mechanical relays and electronic switchboards on my way home. How I hated bricklaying. I was exactly like a high-IQ truck driver who would eventually jackknife an 18-wheeler on the Gulf Freeway.

It wasn't just me under stress. I knew it would happen. Carol decided she'd had enough. She took Brandy and moved in with her parents. It felt like a forced separation, both physically and mentally. I spiraled down the darkest, disorienting, deepest dungeon from which precious light was missing. Months passed. Maybe years. I can't remember enough from those days to describe its wretchedness.

In the end, I crawled up the muddy slope and rejoined the world, holding the hand of a new woman in my life, Ramona Forrester. Ramona was years younger than me, just a girl really. She'd lived a hard-scrabble life herself. She filled the space in my life that Carol could not. In many ways, Ramona's past had been as ragged as mine. She had not finished high school when we met. Her mother was schizophrenic, unstable, in and out of mental institutions, and no good role model for her three daughters. I was giving up on college, and she was planning her future. Maybe this was a life fate had in story for us. Maybe not.

CHAPTER 29

An A-sabbatical

Houston was not good for me. Very little of my time there, at least what I remember of it, pointed me toward any worthwhile goal. Some of my professors took sabbaticals of a year or so to advance their research or retreat to a quiet place and think. My time in Houston was nothing like that. It was more like an a-sabbatical. Still, I did manage to learn something I had dreamed about ever since my brother John and I caught rides in Piper Cubs as kids. I learned to fly. Since pilots must record all flight details in logbooks (dates, destinations, aircraft type, duration, etc.), I can reconstruct some of my faulty memory by associating flights with events. I clearly remember the day I soloed (my first flight without an instructor). By myself. Alone. It can be scary, but finally you have to do it.

Back in those days, there was a quaint tradition that is still practiced today. As soon as a newly soloed pilot lands his plane and steps from the cockpit, he or she is greeted by a group of people, including his instructor, who have been watching his or her "plight." I was no exception.

Instructors never tell students in advance whether a certain day will be their solo day. On my day, I was doing touch-and-goes, as usual. Practicing the perfect landing. Then, without warning.

"Okay, Bob, that's enough. Just taxi back to the hangar and let's call it."

There were some gray clouds building to the west, so I was pretty sure I would not be soloing. Instructors always wait for light winds and calm, clear days for soloing. That was okay with me.

I pulled us up to the tie-down spot where the Cessna 150, N6991S, always parked. I reached over to shut down the engine, but Bill, my instructor, nodded his head and motioned for me to keep the engine running.

"No, no. Go do a couple of touch-and-goes and bring her in after those."

My heart started racing. This was it. Can something be exhilarating and scary at the same time? I promise you – YES. After so many hours of practice, the memory muscle takes over. I performed flawlessly. As I taxied onto the tarmac, a small crowd had gathered. As soon as my feet were on the ground, the onlookers yanked my shirttail from beneath my belt and cut it off with a large pair of scissors. I didn't mind. I was grinning from ear to ear. This was the tradition. As soon as we moved into the airport office, my instructor grabbed a magic marker, wrote my name on my shirt tail, and the date I had soloed. He then climbed a ladder and nailed my patch to the office wall, among dozens of others. That was May 2nd, 1971. My private pilot certificate was issued later that year.

I loved flying. I spent more money renting airplanes to fly than a bricklayer could afford. I simply couldn't get enough. When you are young and wild and not prone to considering dangers, that's when flying can kill you. I came, oh so close, several times. I could fill a memoir with just those stories, but the redundancy would bore you. So, I'll tell only one near-death story here.

The building I'd been working on for a few months was completed at the end of January 1973, a few weeks ahead of schedule. The contractor was so pleased that he'd added a couple of hundred dollars extra to our final paychecks. I had no immediate prospects for more work until the weather improved. Ramona wasn't working either, so I walked in with a big announcement.

"Guess what? We're going to Mexico for a vacation."

She looked at me as if I'd gone crazy, "What do you mean, going to Mexico. We can't do that."

"Oh yes, we can. We all got a bonus for finishing early. We've got spending money."

She smiled, but still didn't believe me, "Mexico? Where in Mexico?"

"I'm thinking Acapulco, on the west coast. I rode a bus down there when I was a teenager, and it's an incredible place."

"Taking a Mexican bus for a vacation does not sound fun. I don't even like taking the bus across Houston."

"Well, we're not taking the bus. We're taking the 140."

"The 140. I don't even know what you're talking about."

"I'm talking about renting the Piper Cherokee out at Hull Field. It wouldn't be that expensive. Besides, the gas prices in Mexico are low. Come on now. It will be fun."

Ramona was only 20 years old and not yet wise to my craziness. A more experienced woman would have stopped this madness with a firm "Absolutely not. I am not going with you to Acapulco in some little airplane that you barely know how to keep in the air."

But that's not what she said. She said, "Lucky us (sarcastically)! I'll start packing."

I drove out to Hull Field to see if the plane would be available. Each rental airplane had a logbook in which qualified pilots scheduled times they wanted to rent the plane. I was lucky enough to sign up for seven days beginning February 4th. The dangers of this foolish act had not yet peep-peep-peeped inside my cracked skull.

Since I had boldly boasted that we would make the dangerous flight, and now the airplane was mine for a week, I needed to do some planning. Hull Field had a collection of aviation maps in the office. I sat down at a big table and unfolded the one for Mexico. Mexico had aviation navigation aids (radio beacons) of the type used in the United States, so that was good. However, the number of beacons in Mexico was far fewer, and they were farther apart. I could use these to guide myself around the country.

After examining the map, what really alarmed me were the mountains. If I flew the most direct route, Tampico to Acapulco, I'd have to cross the Trans-Mexican Volcanic Belt. Two of these volcanoes were well over 17,000 feet tall. The Cherokee lacked the power to climb above 14,500 feet. I'd have to weave my way between snow-capped peaks. Even between the peaks, I'd be at an altitude where supplemental oxygen would be needed, at least by US flight rules. I didn't know the rules in Mexico. This scared me.

I decided that the best route, although considerably longer, would be to fly from Tampico to Veracruz, along the Gulf Coast, then hop over the narrow southern neck of Mexico to the Pacific side. Then, I'd turn northwest and fly up the coastline to Acapulco. I borrowed the map, drove home, and started packing for the trip. Early the next morning, Ramona and I loaded the airplane and started for Mexico. At least, the weather was nice.

Keep in mind that a Piper Cherokee 140 is not a fast airplane; at best, it cruises around 125 mph. And that's air speed, not ground speed. It took us 2.7 hours to get to Brownsville, Texas, where we topped off the fuel tanks. The next leg from Brownsville to Tampico took 2.4 hours.

The leg from Tampico to Veracruz took another 2.5 hours. We didn't stop in Veracruz and spend the night, as we should have. I decided we needed to be on the Pacific side, and I thought we had enough fuel to reach Salina Cruz, a port city on the west coast. We set out across the waist of Mexico, hoping to reach the airport before dark.

Yes, the thought of the high mountains scared me, but I was alarmed at what I was then seeing below. Ramona leaned her head against the window and was half dozing.

"Hey, wake up and look down there. That's some treacherous terrain below us."

She rubbed her eyes and peered down at the rugged greenness below, "Are you sure we came the right way?"

The terrain was solid, impenetrable jungle with pyramids of low mountains jutting from the tangle of vines and trees.

"If we go down in that, they'd never find our bodies."

Since Salina Cruz didn't have a radio beacon, I'd purposely chosen to emerge on the Pacific coast at a point south of the city. That way, I could follow the shoreline northwest and be sure not to miss the airport.

What I didn't anticipate was the headwind we encountered when we turned northwest up the coastline. It slowed our ground speed to a crawl. We were burning what fuel we had in the tanks and making little progress toward Salina Cruz. I was not sure we could make it to the airport. Unlike our course along the Gulf Coast, with endless beaches and barrier islands providing places to land in an emergency, the Pacific Coast had steep cliffs and huge surf pounded boulders at the water's edge: Impossible to safely land.

I reverted to the fuel management training I remembered while learning to fly. The plane carried its fuel in wing tanks. There was a fuel selector that allowed me to draw fuel from either the right or left tank, or from both simultaneously. In low-fuel situations like this, you are taught to switch to drawing fuel from only one tank, either the right or left. Then run that tank completely dry, until you hear the engine sputtering and coughing for gas. It's time to switch to the other tank. This is the best way to manage your diminishing fuel situation. I decided to drain the right tank first and turned the valve. I made a mental note of the time and told Ramona to remember it, too.

We fought the headwind for twenty-two more minutes. Finally, the engine sputtered, and my stomach tightened. I quickly switched to the left tank. This gave me an estimate of how much longer I could stay airborne. About twenty-two minutes. If Salina Cruz didn't appear soon, we'd be in real trouble.

This situation now had Ramona's full attention. No more dozing with her head against the window. We were both counting down the minutes until we'd be forced into an emergency landing, or most likely, a destructive and deadly crash.

After nine minutes of this stress, we began to see signs of civilization. There it was. The port city of Salina Cruz. From a distance, it was all tanks and stacks and rust-colored roofs pressed up against the edge of the Pacific. The Gulf of Tehuantepec was famous for its wind, and even on approach, the Cherokee felt uneasy in the air, nudged sideways by invisible hands. The runway lay in a hot, shimmering basin between the hills and the sea. I remember thinking it was less a destination than a workplace, a port, a refinery, a railroad yard, a town built for fuel and freight instead of tourists.

I did not use my radio to contact the airport. I was fully focused on getting on the ground. When I shut down the engine, the heat wrapped around us. The air smelled of oil and salt. The town felt like a crossing point, not a place to linger. We had no choice but to spend the night there. Our taxi, if you could call a beat-up old post-war Plymouth a taxi, took us to a hotel meant for Mexican and oilfield workers, not tourists. The cost for a night's stay was $2.50!

After a roach-ridden night in Salina Cruz, we topped off the fuel tanks and took off for Acapulco. An uneventful flight. We spent a couple of days there, wandering around town and walking on the beach, but Acapulco was not like I remembered it from my childhood bus trip to Mexico. It was dirty and smelled of rotten papaya. I was ready to head home, and so was Ramona. We sat under a thatched roof on the beach, each sipping a Tequila Sunrise, and planned our escape.

I've never been one who likes retracing steps. So, flying back on the route we came was not an option for me. I don't remember Ramona wanting to do that either.

Our new plan was to fly over the mountains directly to Tampico, the route I avoided when planning the trip. The return home turned out to be equally as harrowing a flight.

The next morning, we stowed our bags behind the seats of the Cherokee and rolled onto the departure runway. I could see the high mountains to my northeast, directly in front of us, and not far away. I started our ascent, carefully adjusting the fuel mixture as I climbed. A few patchy clouds were at 3500 feet, but easy to avoid. Up we went, 3500, 4500, 5500, 6500, 7500, as high as I had flown before. The mountain ridge towered above. The higher I rose, the slower my ascent.

I spotted a pass that looked like the best place to cross into the valley on the other side. I was climbing so slowly that I had to make big circles and rise like a kettle of vultures. Slowly, but finally, I reached 14000 feet and pointed the nose directly at the pass. Situational deterioration didn't creep up on me. It was like slamming into a metal door. I began to feel faint, dizzy, and disoriented. I knew from training what was happening, but I'd never experienced it before. Oxygen deprivation.

As I approached the pass, I noticed boulders and snow just below my wings. The beauty of it trapped my attention. I began to lose control. Evidently, I was not keeping my wings level, and I could hear Ramona complaining, and insisting that I should keep the wings straight. I smiled at the vivid rocks and snow below us. I slumped in the seat and released the yoke. I don't know how long I was out, probably not long, but when the rocks and snow were behind me, I noticed that Ramona had taken the yoke and was guiding it straight and level. I'd let her practice this before, and those lessons paid off. I slowly pushed the yoke forward and began a slow descent into the open air below.

When we were back at 9000 feet, I felt fully in control again. But now, there was another serious problem. We were gliding over a vast carpet of clouds that stretched as far as we could see. The tops of these clouds were 1000 feet below us and probably 5000 feet deep. Although I had had some limited practice with Instrument Flight Rules, I was in no way qualified to make a blind descent through 5000 feet of dense clouds. And what if there were some smaller mountains hidden in the fog below? The insanity of what I was doing in Mexico hit like a hammer.

I don't think Ramona understood the gravity of our problem. She was happy to see me awake and in control. I didn't share with her the turmoil in my soul.

At some point, I resigned myself to the notion that spiraling down through the blinding clouds was our only hope. But when and where should I attempt this? I was pretty sure the clouds didn't reach the ground, but I didn't know that for sure. I tuned in the radio beacon at Tampico. I had a strong signal. That meant I could fly over the cloud tops to the Gulf of Mexico. If I did that, I could get offshore and over the water. Over the Gulf, I would not need to spiral, just make a slow descent, wings level, engine RPM low, flaps down. If the clouds didn't reach the water, I would break out underneath, over the Gulf, and be safe.

Since I am writing this now, you know we ended up safe. But I didn't have to blindly make the descent. It happened that before I got over Tampico, there was a narrow, open tunnel downward through the clouds where I could see the ground. Sometimes called a "sucker hole" by pilots. I took a chance. That hole could have closed in and trapped me in the clouds, but it didn't. We made it through, spent the night in Tampico, and flew back to Houston the next day.

From that day on, I have been a little more cautious, but there were future times when I made similar mistakes and faced similar dangers in the air. This sample of my recklessness will suffice.

CHAPTER 30

Altec Lansing

I emerged from that awful period with one certainty. I needed to go back to San Marcos and try to get back into school. My chances of success were small, but if you'd seen the bottom, as I had, success was measured with tiny divisions on a ruler. If I gained one centimeter (not inches – college people talk metric). If I gained one centimeter for every meter that my classmates strode, that was good enough. I re-enrolled mid-summer of 1974.

I survived Summer II. I retook analytic geometry and passed. Then, I signed up for the fall semester. My course load was heavy again (History, English, and two tough physics courses). Even though Ramona was working, her meager wages would not keep us floating. I must have a job. Heavy financial obligations and no prospects for me working, stress us mightily. It was either a job or dropping out of school again.

All bundled against the cold, I walked down North Guadalupe Street and passed the Palace Theatre in San Marcos, the cramped little movie house that used the pretentious spelling "Theatre." A man was struggling to carry a big amplifier to his car. I offered to help. Once we had the beast in his trunk, he smiled and introduced himself.

"Hey, I'm Frank. Thanks for the help. Let me buy you a cup of coffee and get us out of this wind."

I shook his hand and accepted. We ordered black coffee at a little café a couple of doors down. Frank said he'd just replaced the sound amplifier at the Palace and would ship it back to his headquarters for repair.

He took off his coat and sipped his coffee. "These amps don't fail often, but without sound, the theater has to close. That's big money they lose."

I asked if he lived in San Marcos.

"Oh, hell no. I had to drive from Tulsa. That's almost five hundred miles. Took me a little over seven hours, non-stop."

"Wow. Don't they have anybody in Texas who can fix amps?"

"Altec, that's the company I work for. They have contracts with theaters all over the U.S., broken into territories. The guy who worked in Texas had a baby and quit. The company is looking for a replacement, so I had to deal with this emergency."

"So, you don't have to fix these amps yourself?"

"No, I mostly drive around to a lot of the theaters in Texas and adjust projector sensors. Line things up. Easy work. If an amp goes down, I swap in a spare and ship the old one back." That reminded me of Chester LeBlue's operation.

"Do you think I could do this job?"

"Hell yeah. There's nothing to it. The only bad part is all the driving. You run your own schedule and report once a week. One good thing is they give you a company car, so you don't rack up miles on your own."

"Man. I really need a job like that. I'm in college here and need something for weekends and holidays. I've had some radio experience. Think they might hire me?"

"If you can strip insulation off a wire, they probably will. They're desperate. I sure as hell don't want to drive down here again. Write your name and number down, and I'll send it to the main office."

I got the job. It was almost a miracle, one of the many times I'd gotten lucky against all odds. Altec Lansing flew me to Pennsylvania to pick up my company car, a new Ford Torino. I drove it back to Texas and met the technician who had quit. The company paid him to spend a few days showing me the ropes. I was let loose on several dozen theaters around Texas. My new job brightened my outlook, and I began to look forward instead of slumping into depression.

One cold winter Saturday, I was in deep South Texas, the Rio Grande Valley. It was late at night, near the end of the final showing of some lame movie. One of the ushers from downstairs came to tell me that I had a phone call. There was an extension in the projection booth, and I took the call.

The dynamo roared in the booth, almost drowning out my supervisor's voice on the line. "Benson, how much longer are you gonna be at the Cactus?" I cupped the receiver, straining to hear over the frying hiss of the carbon rods inside the projector.

"They're running the next-to-last reel now. Maybe about an hour and fifteen minutes," I said, glancing at the screen from the projection booth.

"Well, as soon as you finish, I need you to get your butt up to Roswell. The whole sound system has gone down at the Plains Theatre. You've got to get it up before they open tomorrow at noon."

Roswell? That was way the hell northwest, not even in my territory. I asked how far it was.

"Well, I traced the route out on a map, and it looks like it's an 11-hour drive. You'd better leave right now and not wait until you finish up there. Tell the projectionist that you have an emergency and need to leave immediately. Say you'll be back to his routine maintenance next week."

The call ended. I leaned against the console, the projector's hum rattling in my bones. For weeks, my life had been unraveling. This job with Altec Lansing was supposed to provide me with easy-going income, just weekends and holidays. Mostly, that was true. But it left me no time to study or finish assignments. By February, I was way behind my classmates. All this driving left me too exhausted to drag myself into class. I stopped showing up on Mondays, and my professors noticed. I was losing weight, losing health, losing ground. Devastation.

Now Roswell? I must be in Roswell before noon tomorrow. I stared at the carbon arc, hissing and spitting light, and knew something had to give. Now I had to drive all the way to Roswell. Even though the weather was nice here in the Rio Grande Valley, I'd heard on the radio that there was a significant ice storm blasting its way through the Panhandle.

"Hey Jackson, that was my boss calling." Jackson, whose last name I don't remember, turned to hear what I was saying, keeping one eye on the screen, looking for the first white dot that would prompt him. The present reel was almost empty. In a couple of minutes, he'd need to switch to projector two, which held the next reel of the movie. Projecting a movie required from five to seven reels of 35mm film, each 15 minutes long. That was the maximum time the flaring, burning, and blasting carbon-arc rods could last. His job was to prepare a second projector with new carbon rods and mount the next reel before the other projector ran out of light.

Jackson turned his head toward me, swinging his dangling blonde ponytail off his shoulder. He kept an eye on the screen. "I couldn't hear you. What did you say?"

"I said that one of our sound systems in New Mexico has gone down, so I'll have to head up there right now.

Your system's doing okay. It'll be fine 'til I get back down here next week."

He nodded once, barely, the way men do when they've already filed the information away and decided it doesn't change what their hands need to do in the next minute. He was standing between the two projectors, sleeves rolled, tie loosened, his body angled toward the running machine as if listening to it breathe.

The booth smelled the way projection booths always did then. Hot metal. Burnt carbon dust. Ozone from the arc. A faint, sweet undertone of grease that had soaked into the wood decades earlier and would never come out. The floor vibrated slightly underfoot, a constant tremor from the gears turning at a speed that never varied.

On the screen below, the picture rolled on, steady and bright. Somewhere inside the machine, a reel was nearing its end.

Jackson leaned closer to the porthole. He didn't stare straight through it. He watched the corner.

The first white dot appeared.

It was small, easy to miss if you didn't know to look for it, tucked into the upper right of the frame. To the audience, it meant nothing. To Jackson, it meant everything. He reached back without looking and flipped the motor switch on the second projector.

The dormant machine came alive with a low whine, belts catching, speed climbing. He struck the arc, bringing the carbon rods together with practiced ease. A hiss, then the hard, steady light bloomed. He adjusted the rods by feel, fingers addled from years of it, watching the ammeter needle settle where it belonged.

Seven seconds.

He didn't count them. He didn't have to. His body knew the distance between the dots the way a musician knows the space between notes.

The second dot appeared.

Jackson threw the changeover switch in one smooth motion, a gesture so quick and sure it barely registered. The picture on the screen never faltered. No jump. No flash. No sound. For the people in the seats below, the movie continued, as if it had always been a single unbroken film.

Jackson closed the dowser on the first projector and shut down its arc. The light snapped off. The machine spun down, coasting, relieved of duty. He moved to it immediately, rewinding the spent reel, hands moving with the same care he'd used to start the next one.

Only then did he turn to me.

"You gotta go," he said, not asking.

I nodded.

He didn't seem annoyed about the sound system in New Mexico going down. He didn't ask when I'd be back. Emergencies were part of the job. The booth had taught him that. Everything ran until it suddenly didn't, and when it did, someone had to move fast and quietly and know precisely what mattered in that moment.

I stood there a little longer, watching the film wind through his hands, watching him prepare for a future interruption that everyone else would never notice. I grabbed my toolbox and waved as I started down the narrow steps to the lobby below. I was soon dodging tipsy Saturday-night drivers and headed north. Eleven hours would give me time to consider my options. The truth was that I didn't have any options. Without my job with Altec, I couldn't pay the rent at our old house on West San Antonio Street. Then there was Tracy's child support. A job in San Marcos, if one could be had, paid almost nothing. Not an option.

Working for Altec was the kind of work that meant you slept when you could and drove when you had to. Sound systems did not care about holidays or geography, and neither did the people who owned theaters. If the system went down during a weekend run, especially in a small town where the theater was the community's living room, it was treated like a medical emergency.

It didn't matter that I was in Pharr, Texas, down in the Rio Grande Valley, finishing a late-night adjustment on a system. My supervisor had said that Skeet Noret's theater in Roswell, New Mexico, had lost sound. Noret owned several theaters, including the Movieland in Lamesa, and when his theaters went dark or silent, it mattered. My supervisor didn't ask whether I could go. He told me to stop what I was doing, pack up, and start driving north, first to Lamesa. Meet Skeet. Then on to Roswell.

I remember stepping outside after the call. The Valley air was still warm, damp even at that hour. It felt strange knowing that by morning, I'd be crossing country where the temperature could drop below freezing. I loaded my tools, took one last look at the Pharr marquee glowing against the dark, and pointed the car north.

Lamesa was quiet when I arrived. Skeet was waiting. He was calm, practical, and already thinking ahead. He asked how fast I could get to Roswell in my company car. I gave him an honest estimate. It wasn't slow, but it wasn't fast enough for his liking.

Then he did something I hadn't expected. He offered me his Datsun 240Z. I had seen his car before. Everyone in the Texas Panhandle had. In 1970, a Japanese sports car in West Texas might as well have been a spaceship. It sat low, long hood, short rear deck, nothing flashy, but purposeful. It didn't look fast in a cartoon way. It looked fast the way machinery does when it's built to do one thing well.

Skeet handed me the keys without ceremony, no warning speech. No conditions. Just said, "You'll make better time in this."

He was right. The 240Z felt different the moment I pulled out. The steering was tight, the engine smooth and eager. It didn't roar like the American muscle cars people were used to. It pulled. Cleanly. The faster it went, the more settled it felt, like it wanted to live at speed.

Once I was clear of town, and on the open road, the landscape flattened out in every direction, long straight highways. No traffic. No lights. Just the faint rise and fall of the land and the beam of the headlights stretching far ahead.

I eased the car up to ninety, then higher. At a hundred plus miles an hour, the engine wasn't screaming. It was working, but comfortably, a steady mechanical hum that felt intentional rather than strained. The speedometer needle stayed planted. The car tracked straight. The road came to me faster than I had ever known before.

At that speed, distance collapses. Towns lose their meaning. You're no longer traveling through the land so much as skimming it. I wasn't weaving or pushing the car beyond its limits. I was letting it do what it was designed to do, and for long stretches I held that speed, confident in the machine and in the emptiness of the road.

At first, it didn't feel reckless. It felt efficient. But somewhere west of Brownwood, the air changed. I didn't notice it right away. The road looked the same. The sky was clear. But winter announces itself quietly in that part of the country. A slight sheen on the asphalt. A patch of road that reflects light differently than it should.

I hit the ice before I knew it was there. The rear of the car stepped out so suddenly that there was no time to correct. No gradual warning. One moment, the steering wheel was solid in my hands, the next it was meaningless. The car rotated, slow enough for me to understand what was happening, fast enough that I couldn't stop it.

There is a particular silence that happens when a car leaves the road at speed. The engine noise drops away, replaced by wind and the dull thudding of tires over uneven ground. The headlights swung wildly, then pointed at nothing useful at all.

I remember thinking, very clearly, that this was how people died. Not dramatically. Not with fire. Just a miscalculation, a patch of ice, a body flung against something solid. But there was nothing solid to hit.

The land out there between Brownwood and Roswell is vast and forgiving. Flat. Empty. No trees close to the road. No ditches deep enough to trap a spinning car. The 240Z slid, bounced, spun again, and then slowed, shedding speed across open ground until it finally came to rest, angled wrong but upright.

I sat there for a long moment, hands still on the wheel, heart hammering, the smell of cold air filling the car. When I finally stepped out, my legs shook enough that I had to lean against the fender. The car wasn't damaged. Not a crease. Not a broken light. The tires were dirty, the undercarriage dusty, but the machine had survived intact - so had I.

I stood there alone in the dark, the realization settling in slowly. I had been traveling faster than most people ever do, in the middle of the night, in winter, far from anyone who could help. The car had given me speed, but the land had shown mercy.

After a while, I got back in, turned the key, and the engine started as if nothing had happened. I drove the rest of the way more carefully. The urgency was still there, but it no longer owned me. When I finally reached Roswell and fixed the sound system, no one asked about the drive. They only cared that the movie could be heard again.

That night, that car, and the fine line between efficiency and disaster will always stay with me. The Datsun 240Z had been fast, yes. Fast enough to make time bend. Fast enough to make me believe I was ahead of everything. This was one of the many times in my young life that I barely escaped death.

I drove back to Lamesa in a terrible mental funk. Everything I was facing seemed destined to defeat me. My dreams of college seemed vanishingly distant. Even with my Altec job, I had run out of money. Child support was looming, and I could be dragged away to jail if I didn't pay. I knew that because Rachel would not let me forget it. I didn't know how to make a difference. I dropped off the Datson in Lamesa and searched around for a pay phone. I needed to call Ramona and tell her what happened, and when I might be home. I remembered what my father had said about truck drivers. I almost wished my IQ was 80, so my mind would not keep fantasizing about changing my life.

CHAPTER 31

Lead Bricks

Even with everything pressing down on me, I was still in classes and making it through.

The physics faculty at SWT may not have been unique in the liberty they allowed their students, but the freedom we had was breathtaking. Yes, we had textbooks, and yes, we learned mathematics and solved problems, but there was a hands-on, get-your-brain-and-body-dirty atmosphere. We touched things that would kill you. We recreated the ground-breaking experiments and discoveries that had transported us from the dark ages to our present state of knowledge. We rolled steel balls down inclined planes, we built sound amplifiers from scratch, we vaporized mystery powder in flaring carbon rods (just like the ones in movie projectors) to expose a spectrum on 35mm film, we recreated the Millican Oil Drop

Experiment that had first measured the charge on an electron, and we worked with radioactive isotopes that could damage your DNA at the end of two-foot-long tongs.

Outside the science building, sitting on a concrete slab, by the sidewalk, sat a unique structure. It was made of bricks in the form of a perfect cube, measuring six feet on any side. There were no windows or walk-in doors in the building. But there was a heavy iron door about the size of a chessboard, secured with a sturdy padlock. Students and faculty walked by this blockhouse every day, and few knew what was inside.

I knew what was in there, although I'd never seen it.

Did I mention that the brick blockhouse was not made of fired clay bricks, but of solid lead bricks, each weighing 26 pounds? Inside the blockhouse was a powerful, extremely dangerous radioactive neutron source consisting of Americium-241 and Beryllium-9.

Am-241 and Be-9, when mixed, deliver a harsh dose of neutrons that penetrate your clothes, your skin, your organs, and cause mutations and cancer. Anyone who wanted could have sneaked to the blockhouse in the middle of the night with only a crowbar and carried the dangerous radiation away.

Finally, it came time for my class to use the source for an experiment. The objective was to measure the half-life of certain silver isotopes. Mr. Spear had gathered the class in a preparatory meeting before we began dealing with the radiation.

Arthur Spear was not one of those professors who tried to dazzle you with brilliance or personality. He seemed built for endurance, not display. He taught the labs and the bread-and-butter courses, the unglamorous parts of physics where wires had to be connected properly and measurements had to be taken twice because the first answer was probably wrong. Correctness mattered to him more than inspiration. He simply showed up, semester after semester, and made sure the experiments worked and the students survived them.

He called the class to order, "Now, everyone, please pay great attention to what I am about to explain. The experiment for measuring the half-life of silver isotopes can be a little dangerous if you don't correctly follow the rules. You've all seen the lead blockhouse outside the science building. Inside that house, there is a strong radioactive source of penetrating neutrons. Today, two of you will be in charge of opening that blockhouse and facing that source. The remainder of the class will have a role in this, too."

He had our attention. He opened a drawer in the desk in front of the classroom and retrieved a small white cotton bag with a drawstring. I recognized it as a Bull Durham tobacco bag, from which my friends and I had secretly rolled cigarettes as kids. He opened the bag and poured the contents onto the desktop. The coins made a bright, ringing sound, with a higher pitch than nickels or pennies.

"Now these are silver dimes I've saved since they were taken out of circulation about a decade ago. They are not pure silver, even though they are called so. They are 90% silver, and they will be good enough for what you will be doing. I am picking only two students to directly handle the dimes during the experiment. But the rest of you will have a role."

The class looked around to guess who might be selected to be exposed. Since I was not one of the better students, I didn't expect to be chosen. In Mr. Spear's understated manner, he said, "I am choosing Benson and Franklin to do the dangerous part, because they are older."

My heart fluttered. Clearly, I was the oldest student, but George Franklin didn't seem much older than the other students in the class. I wondered if we were chosen because we might die first from causes other than radiation poisoning.

"Now, naturally occurring silver is made up of two non-radioactive isotopes, Ag-107 and Ag-109, in about equal measures. When this bag of dimes is placed next to the hot source in the blockhouse, the Ag-107 picks up an extra neutron and becomes Ag-108, which is unstable. The isotope decays very fast with a half-life of 2.4 minutes. That means these dimes, when removed from the blockhouse, will lose their radioactivity very quickly. You are going to measure that decay with the multichannel analyzer down the hall. Suppose you're all disciplined and fast and can get the dimes out of the blockhouse and into the analyzer in around two and a half minutes, already your dimes will have lost half of their activity."

The class looked nervous. Mr. Spear opened another drawer and retrieved a strange-looking pair of tongs.

"The 'hot' bag of dimes will be handled using this pair of tongs. George will use them to deposit the dimes into the blockhouse, where we will leave them for five minutes. Bob will retrieve the irradiated dimes and run them from outside and up to the analyzer."

Mr. Spear scooped up the scattered dimes and put them back in the bag. He handed the bag to George along with the keys to the blockhouse and the tongs.

"Now, George, after you unlock the blockhouse door and open it, you'll see a little platform inside. Use these tongs to place the dimes on that platform. Do it carefully and deliberately, but don't dally around. Try to have the door open, the dimes placed, and the door closed as swiftly as you can. But pay attention and do things safely."

George nervously nodded his head.

"Now, when that door is closed, you pass the tongs to Bob, then start your stopwatch. Let the dimes stay for five minutes."

So far, this sounded simple. Mr. Spear told me to look over George's shoulder and note the spot where the dimes are placed, so I wouldn't waste time retrieving them after five minutes, when the dimes were hottest. They start decaying fast.

You'd think these are the sort of statements that would make a professor smile. But no. Arthur Spear was a stone-faced man that I'd never seen smile at anything. As they say, he was as serious as a heart attack.

"Now, here is what the rest of you will be doing. I want one of you posted at the door nearest the blockhouse, and someone else at the door where the multichannel scalar is. You doormen. You will open the doors, so Bob doesn't have to risk dropping the hot dimes on his way to the analyzer. The rest of you should scatter along the hall and stairs to alert other students to stay out of the way, as Bob hastens along with his radioactivity."

Mr. Spear asked George and me to repeat back our instructions, and we did. He waved all of us out of the room and to our assigned tasks. We all did our jobs perfectly. We measured the half-life of Ag-108, and our results were quite close to the curve shown in textbooks.

What I most clearly remember is that while I was holding those irradiated silver dimes with those two-foot-long tongs, I could feel a tingling in my arm that, at the time, I was sure was caused by the beta decay (electrons) and gamma rays (photons) blasting from the Bull Durham bag. I now understand that my perceived sensation was completely in my imagination.

CHAPTER 32

Broken Bones

SWT had a policy called Credit by Examination, or Advanced Standing. A student could earn credit for a course by scoring an A or B on a special exam prepared by the professor, usually more comprehensive than a final. I used it twice. I'd been hired as the Astronomy Lab Instructor for a semester, learning the material by staying one step ahead of the class, and afterward I took the Advanced Standing exam and collected four hours of credit. I found a senior-level Flight Instruction course whose syllabus matched the ground school material from my private pilot certificate almost exactly. I tested out of that one too. Seven hours of credit, and my graduation date moved closer.

When the astronomy course ended, so did my small stipend from the physics department. I was walking down the hallway when Dr. Crawford stopped me.

"Hey Robert." I almost kept walking. Almost everyone called me Bob.

"I hear you might be looking to continue your stipend. Is that right?"

"Yes, sir. Dr. Anderson said the department wouldn't need me until Astronomy runs again. That little bit of money was sure helping me."

"Well, I've got a research grant with some money for student help. I could pick you up for a semester. We'd work around your classes."

I'd taken one course from Crawford but had never been inside his lab. I knew he worked with lasers. That was the sum of what I knew.

"Come take a look at what I'm doing there," he said.

He unlocked the door. All the windows were covered with heavy black paper. The room smelled of chemicals and cool metal. In the center sat tables as large as pool tables, made of thick steel.

"Interesting tables," I said.

"Isolation tables. A lot of what I do involves making holograms. If the laser shakes even one wavelength of light, the experiment fails, it can't tell a vibration from a moving object." He flipped a switch and I heard a soft exhale of compressed air. He tapped one of the legs with his knuckle. "Air springs. The tabletop rides on compressed air so the building's vibrations never reach the optics. Think of it like a boat on water. Waves pass underneath, but the deck stays still. Without this table, all you'd photograph is noise."

"What do you photograph with it?"

"Here's the trick." He walked me to the bench. "We coat a glass plate with light-sensitive material and put an object in the spread-out laser beam, say, a hot cup of coffee. We take the first exposure. Then I turn off the laser, wait fifteen seconds, and take a second exposure on the same plate. By then the coffee has cooled slightly, and the cup is a hair smaller than it was. Things expand when heated, contract when cooled."

He opened a drawer and unwrapped a developed glass plate.

"Hold this up to the light."

I held it up. There was the cup, covered in zebra stripes, bands of dark and light running across the surface of the glass like the rings inside a tree.

"That's an interference pattern. Those stripes let me measure changes in the size of the cooling cup smaller than the wavelength of red light. Before lasers, this level of precision was nearly impossible."

It was about the most beautiful thing I had ever seen made by science. Not the image itself, though that was striking, but the fact that something invisible, the shrinking of a cup as it cooled, had been made into a picture you could hold in your hands.

Crawford kept walking. At the back of the lab was another door.

He opened it, and I stopped.

The room held fifty white rabbits, each in a small cage, most of them pulling at alfalfa hay. Peaceful animals. Calm, pink-eyed, going about their business. Several had bandaged hind legs.

Crawford let me look for a moment before he spoke.

"Part of the job is keeping these animals fed, watered, and clean. Cages need daily attention. Waste goes down to the dumpster."

"What are they for?" I asked. I already had some ideas I didn't want confirmed.

"Without them, I couldn't do this research, and I couldn't offer you a job." He paused. "I have to break their legs to run my experiments."

He let that land.

"I don't talk about this much. But I get the feeling you've spent some time around farm animals. I don't think you'll have a problem with it."

I wished he hadn't said that. I knew what he meant by it. My accent gave me away. My clothes, which did in fact come from Goodwill, gave me away. He hadn't hired me because he detected a scientist. He'd hired me because he needed someone who could manage blood and broken bones without flinching.

I stood there looking at the rabbits eating their hay.

He was right and wrong at the same time. He was right that I'd seen things, the turkey massacre alone would have sorted that question. But I'd never stopped flinching. I fainted around blood. I always had. Standing in that room, with the bandaged legs and the smell of hay and animal, I felt the distance between where I was and where I'd come from collapse to nothing. I thought I'd been building something new. It turned out I was still the same boy.

I needed the money. I took the job.

The science, once I got past the rest of it, was genuinely worth knowing. Bone tissue is piezoelectric, bend or stress a bone and it produces small electrical charges, negative where compressed, positive under tension. New bone grows where the charge is negative. That suggested electricity was part of the body's own repair signal, and Crawford was testing whether artificial current could accelerate healing in a fractured bone. The laser measured the response. The same glass plates, the same interference patterns, the same zebra stripes, now measuring the slow, invisible work of a body knitting itself back together.

I learned everything in that lab. I could set up the lasers, take the double exposures, develop the plates, run the darkroom.

And the longer I worked there, the more I kept turning a question over. This whole lab, every instrument in it, was a sophisticated ruler. We measured things. Tiny things, yes, but that was all we were doing: measuring.

One afternoon I was soldering a circuit board at the bench, the kind of board common in 1975, components with long metal legs pushed through drilled holes, wide copper traces you could follow with your eye, and I thought: if we can see a coffee cup shrink as it cools, why can't we see a resistor change size when it fails?

I built a test board with three resistors. I ran Crawford's double-exposure method on it, one image of the functioning board, one after I'd killed a component. The interference pattern showed exactly which resistor had changed. The faulty part announced itself in stripes.

Crawford and I wrote it up. Applied Optics published it in January 1976.

We had a published scientific paper, and my name was on it, too. It still seemed like something that had happened to a different person.

I graduated that spring with a GPA just under 3.0. Nothing to brag about. I applied to the graduate physics program at Texas A&M and was accepted for the fall semester of 1976.

I didn't get in on my grades. I knew that.

I got in on a glass plate covered in zebra stripes, and on fifty rabbits I cleaned up after every morning, and on the stubborn habit my mind had of asking what two unrelated things might have in common.

I was about to find out what a real physics program looked like.

CHAPTER 33

Cosmic Rays

When I drove into College Station in the fall of 1976, I remember first noticing how wide everything felt. The roads were wider than San Marcos. The parking lots stretched longer than seemed necessary. Even the sky felt larger, or maybe that was just the land around it, open, flat, disciplined. Nothing of the loose river-town feel I had known at Southwest Texas State. No limestone hills leaning in. No familiar curves in the road. It felt engineered.

The campus seemed built to endure, not to charm. Long brick buildings squared off and orderly. Broad lawns trimmed close. Sidewalks that ran straight as if laid with a ruler. There was seriousness to the place. Even the air felt purposeful.

I had applied, in a moment of what I can only call boldness, to three graduate programs: Cornell, Purdue, and Texas A&M. Cornell passed. Purdue and A&M both said yes. I chose A&M because it was in Texas, and Texas was what I knew.

At Southwest Texas State, I had felt comfortable in my skin. I knew the classrooms. I knew the cadence of the professors' voices. I knew how I stacked up against the other physics majors, and I stacked up well. At A&M, I felt out of place before I ever stepped into my first classroom.

Graduate orientation was held in a lecture hall that seemed twice as large as any room I had studied in. The other students were already there when I arrived. Most sat quietly, flipping through folders. No one talked much. They carried themselves with a kind of composure that was not arrogance but something close to settled certainty. They looked like people who had already decided something about themselves.

I began listening to where they had come from. Caltech. Illinois. Berkeley. Rice. They spoke easily about undergraduate research, advisors, and labs they had worked in as juniors. I worked hard at Southwest Texas State. I had solved the problems. I did well. But I had not come from their world, and standing in that room, I knew it.

The physics building itself felt colder than the rest of campus. Fluorescent lights humming. Chalkboards running the full length of the wall, equations ghosted across them from the day before. One professor began a discussion as though we had all been thinking about the same problem for years. He spoke quickly, not to impress, but because that was his natural speed. Heads around me nodded. Mine did not.

At Southwest Texas, I had raised my hand often. Here, I waited. I listened. I wrote down every symbol and told myself I would catch up later.

There is something humbling about being surrounded by people who are not just bright but intensely serious about being bright. These were not students drifting through on charm or momentum. They had been chosen. Hand-selected from the strongest physics departments in the country. It was as if the field had narrowed and I had slipped through the last gate just before it closed.

I walked back to my car that first afternoon with a strange mixture of pride and unease. I wanted to be here. I had worked to be here. And now that I was, I understood something I hadn't fully considered before: admission was not arrival. It was only permission to try.

The first task was to find a faculty mentor. That was not written in any handbook, but it was the first real test. Every graduate student had to be accepted by a tenured professor who would chair their committee and direct their research. You spent the early weeks meeting faculty, reading their work, trying to determine whose lab might take you in. No mentor meant no path forward.

There was one astronomer on the faculty, Dr. Andrew Young, who seemed like the right match. Unfortunately, he was only a visiting professor and could not chair committees. He suggested I go talk to Nelson Duller, a cosmic ray physicist, a man who did astronomy with subatomic particles rather than photons. That suggestion turned out to be some of the best advice anyone ever gave me. Without Nelson Duller, my first semester at Texas A&M might have been my last.

Nelson Duller was a quiet man. Not distant. Just quiet. He did not fill space unless there was something worth saying. When I walked into his lab, I would often find him seated at his typewriter, shoulders slightly forward, fingers moving with a steady mechanical rhythm. The sound of the keys striking paper was as much a part of that room as the smell of chalk and machine oil.

Every single day I knew him, he typed. Page after page. I never asked what he was writing. I assumed it had something to do with physics, calculations, notes, observations. But there was a discipline to it that went beyond lab work. It was daily. Unbroken. It was his journal.

He encouraged questions. That mattered. In a department where the air itself felt competitive, he never made a student feel foolish for not knowing something. If I didn't understand a concept, he would pause, turn slightly in his chair, and walk me through without impatience. No performance in him. No testing egos. Just explanations.

He praised me, too. Not lavishly. Not in ways that would make a man swell. But directly. Enough to let me know he thought I belonged there. That I was capable of the work. That I was not just surviving.

In graduate school, belief can wobble. There are days when you measure yourself against the room and come up smaller than the day before. Dr. Duller did not rescue me from those moments. He simply treated me as though I was already the physicist I was trying to become.

I remember one of the first times I needed his help. I went into the lab quietly. He was typing. He finished the line he was on before looking up.

"Yes?"

"I'm supposed to teach the ESR lab next week," I said. "I've read through it. I understand the equations. I've set up the equipment and gone through the procedures, but I'm not sure I understand what I'm really seeing."

He nodded once and pulled the paper from the typewriter. He set it squarely on the stack beside him, aligned the edges with care, then turned fully toward me.

"Good," he said.

I waited.

"If you think you understand it completely, you probably don't."

That eased something in me.

He stood and walked to the bench where the Helmholtz coils sat, still half-wired the way I'd set up the experiment. He picked up a piece of chalk and moved to the board without hurrying.

"What do the students think they'll be measuring?" he asked.

"The spin of the electron," I said. "Up and down states. The resonance condition."

He wrote the energy equation on the board, the gap between levels expressed in terms of the magnetic field strength.

"And what does this mean?" he asked.

"The magnetic field splits the energy levels," I said.

"Yes." He underlined the field term. "The field does not change the electron. It changes the energy landscape."

He turned back to me. "The electrons already have spin. They already have a magnetic moment. The field simply makes the two orientations cost different amounts of energy."

He drew two horizontal lines on the board, one slightly above the other.

"This one," he said, tapping the upper line, "is slightly more expensive."

"And the microwaves," I said, "supply the energy difference."

"Only at the right frequency," he said. "Resonance is not magic. It is bookkeeping."

He stepped away from the board and walked to the coils. He rested his hand lightly on the frame.

"The students are not seeing an electron flip. They are seeing absorption. Many electrons, responding together."

He looked at me, not testing, just steady.

"You do not need to understand every part of the microwave cavity to teach this. You need to understand the story."

"The story?"

"Yes. There are two states. A strong magnetic field separates them. Microwaves match the gap. The system absorbs energy when the bookkeeping is correct. When you sweep the field slowly, you move the levels past the fixed microwave frequency. At one field strength, the numbers match. That is the peak they see."

I nodded. The structure settled into place.

"They'll want to know what spin really is," I said.

He gave a slight half-smile. "So will you."

He walked back to his desk and slid a fresh sheet of paper into the typewriter.

"You are not expected to know everything," he said, adjusting the carriage. "You are expected to think carefully."

The keys began again.

"And when they ask questions?" I asked.

He did not stop typing.

"Encourage them," he said. "If you do not know the answer, say so. Then find it."

The room returned to the steady rhythm of metal on paper. I stood there a moment longer, looking at the board, the two lines, the small gap between them.

"Thank you," I said.

He nodded once, still typing.

That was how he taught. Not by lowering the material. Not by dramatizing it. Just by placing the pieces in order and trusting that if you stood there long enough, they would begin to make sense.

This was the man who would determine my fate.

CHAPTER 34

The Brown Houses

When we first moved to College Station, I was old enough to know better and young enough to think I could work out anything that came to me. It was me, Ramona, and our son Eric. Eric was still a toddler, small enough to fit on one hip, loud enough to fill a room. I had come to Texas A&M to survive graduate school. That is what I told people. I did not yet understand that survival meant more than passing classes and building an experiment. It meant holding together a family while the ground under you shifted.

We rented a small two-bedroom apartment first, the kind with thin walls and appliances that hummed louder than they cooled. When an opening came up in Texas A&M's Married Student Housing, what everyone called the Brown houses, we moved as soon as we could.

The Brown houses were plain, uniform buildings, brick the color of dry dirt, set in rows that looked almost military. But inside that sameness was a world I had never lived in. Heavy with foreign student families. In the early evenings, as the light softened and the air cooled just enough to breathe, you could walk between the buildings and catch the smells of dinners I had never encountered growing up. Indian food rich with spices I could not name. Arab dishes thick with garlic and something sweet and sharp at the same time. Asian meals that steamed and hissed and carried scents of ginger and soy. Those smells drifted through open windows and under doors. At the time, I did not think of it as culture. I thought of it as life going on around me while I tried to keep my own piece of it from falling apart.

Ramona worked on campus as a secretary. On top of that, she typed dissertations and master's theses for students nearing graduation. In those days, typing meant real typing. No word processors to clean up mistakes. No easy revisions. She sat at a typewriter and turned stacks of handwritten pages into something that could pass a committee's inspection. She did this while raising a toddler and keeping our small apartment from sinking into chaos.

I was in the physics building most days and many nights. I did not have proper funding for what I hoped would become the foundation of my dissertation, assuming I made it that far without failing out. That word hovered over me more than I admitted. Failing. It was not just academic. It felt personal.

Dr. Duller suggested I work on detecting cosmic rays, muons specifically, those particles born from collisions of high-speed protons with atoms high in the atmosphere. They rain down on us all the time. Invisible, constant, indifferent. He told me he had once built a Geiger counter from old burned-out fluorescent light bulbs. That idea stuck with me. It sounded like something a man could do with his hands. It sounded possible.

The process was simple in outline and complicated in practice. Take a four-foot fluorescent bulb. Use a car battery and a length of tungsten wire to snap off the ends cleanly. I would connect one end of the tungsten wire to the battery, wrap a loop around the metal connector at one end of the tube, then touch the other end of the wire to the other battery terminal with pliers. The wire would glow red-hot almost instantly. The glass would give way in a clean break, separating the metal prongs from the long tube.

Once both ends were off, I used a bottle brush to scrub the white phosphor powder from the inside surface. What remained was a long clear glass tube. From there, I machined plastic plugs with holes for gas and wire. The outside of the tube was wrapped in aluminum foil. A wire ran down the center. Helium gas flowed through the tube. With high voltage between the foil and the central wire, it became a Geiger tube capable of detecting muons.

Each time a muon passed through, there would be a pulse. That pulse could be measured. That measurement could be data. Data could become a dissertation. A dissertation could become a degree. A degree could justify all of it.

The helium flowed continuously, entering one end and exiting through a small poly tube at the other. To prevent room air from flowing back in, I set up a simple backflow prevention system. The exiting poly tube ran into a flask half filled with water, sealed with a double-hole rubber stopper. One hole for the helium to bubble out. The other to equalize pressure.

It was a setup that worked. Until it did not.

Michael, our second son, was born in December 1977. I remember holding him and feeling both joy and a weight I could not name. Two children. No money to speak of. A project that depended on salvaged fluorescent tubes and borrowed equipment. Something pressed down on me in those months, heavier than fatigue, heavier than the usual worry. I did not have a word for it then. I am not prone to depression, but I was close to it there, in that small apartment, with the bills and the experiment and the thing I needed to prove to myself.

One day I left the Geiger setup on a lab bench overnight. That evening, a Texas Norther came through College Station. The kind that drops the temperature fast and shifts the air pressure without asking permission. Sometime during the night, the change in atmospheric pressure pulled the rubber stopper down into the throat of the flask. It wedged tight.

The next day I found it that way. I could not pull it free. I tried twisting. I tried prying. Nothing worked.

That is where the lesson begins.

I got the idea, a foolish and dangerous one, that I would pump compressed air through the poly tubing into the flask and hold my thumb over the other hole in the stopper. The pressure would build and force the stopper out. It sounded reasonable in the shallow part of my mind. It was a shortcut. It would save time. It would prove I could solve a problem without asking for help.

I did not consider the tensile strength of glass. I did not calculate pressure. I did not think about what happens when a sealed container made of brittle material is forced past its limits.

I held the flask at chest level.

I remember the instant before the explosion as ordinary. That is what stays with me. There was no warning. No creak. No visible crack spreading. Just a moment of intention.

Then the flask exploded with an ear-shattering boom that could be heard down in the lobby of the building. Not pop. A blast. The glass did not break. It disintegrated.

Shards blasted into my chest and arms. The sound alone stunned me. I went completely deaf, as if someone had clapped their hands over both ears from the inside. The world narrowed. I almost lost consciousness.

When I could see clearly again, I realized none of the glass shards had struck my eyes. That fact stayed with me more than any equation I learned there. If a fragment had been a fraction of an inch higher, my life would have bent in a different direction.

As I staggered, I saw blood trailing from my right arm. Not a little. A lot. One fragment had severed a major artery. The blood was not oozing. It was leaving.

A student working in a lab on the floor below heard the blast and came running. He wrapped a rag around my arm and tightened it. He called an ambulance. I sat down because the room would not hold still. I thought, not about death, but about Ramona. About Eric and Michael. About how thin the margin had been between a bad idea and something final.

At the hospital they stitched me back together. I did not need a transfusion. That is what they told me.

I was back in the lab a couple of weeks later. The custodians had cleaned up as best they could. Still, under the benches and in the cracks of the floor, tiny fragments of Pyrex remained. I am certain that if someone searched carefully even today, they would find pieces from that flask. Evidence of a moment when I mistook impatience for ingenuity.

The responsibilities waiting at the Brown houses had not changed. Ramona still typed late into the night. Eric still ran through the apartment with more energy than the walls could hold. Michael still woke up crying in the dark.

The smells of evening meals still drifted in through the windows.

But something had been settled. Not cured. Settled. You cannot force what will not move. You cannot hold a sealed thing past its limits and expect it to yield politely. I had known that about materials for years. I had not yet applied it to myself.

I walked away with a scar and my sight. That was more grace than my decision deserved.

And I learned that in both physics and life, pressure builds quietly until it does not.

CHAPTER 35

Zombies in the Night

When I arrived at Texas A&M in the late nineteen seventies, I thought I knew something about computers. At Southwest Texas State I had worked on a PDP-11. You could touch it. You could sit at a terminal and feel like you were in conversation with the machine. It responded in something close to real time. You typed, it answered. If you made a mistake, it told you quickly enough that you did not forget what you had just done.

The IBM 360 at A&M was another creature altogether. It did not speak to us. It did not see us. It did not even know we existed.

The 360 lived in a walled room deep inside the computing center. You could not simply walk in and lay hands on it. There were chosen people who were allowed near it. They wore badges and carried themselves with the mild authority of temple attendants. We graduate students stood outside the wall like petitioners. If the PDP-11 had been a tool, the 360 was a god, and it demanded ritual.

The ritual began with punch cards: manila-colored IBM eighty-column punch cards, stiff as thin bone, each one holding a single line of FORTRAN code. If you dropped a card and bent a corner, you might as well tear it up and re-punch it. If you dropped a stack and got them out of order, you felt your stomach sink in a way that had nothing to do with hunger.

My project did not have grant money. Many of us did not. Professors with funding could pay for daylight machine time. Their jobs ran in the mornings and afternoons. The rest of us waited until midnight, when computing was free.

There was something ghostlike about the way we unwashed and sleepy-eyed students appeared on campus after eleven. The daytime students were gone. The sidewalks were empty and quiet except for the hum of sodium streetlights, the soft grind of our little two-wheeled dollies stacked with punch cards, and the shock of ear-damaging explosions going off every few minutes.

The university hated Great-tailed Grackles, who found it desirable to roost in the oak trees lining the campus sidewalks. As soon as it turned dark, pickups with propane guns mounted upright in their beds would start their rounds, setting off ear-shattering blasts you could hear for miles. The startled grackles would scatter from the trees and move to a different spot fifty yards away, still directly over a sidewalk. When the popguns proved useless, they hired falconers using Red-tailed Hawks and Great Horned Owls to move the birds around. That did not work well either, and they finally stopped after a Red-tailed Hawk scooped up and ate an elderly woman's miniature chihuahua right off the end of its leash. Texas A&M could engineer almost anything, but they could not engineer a method to rid the campus of grackles.

Those dollies were built to carry the gray cardboard boxes that held our programs. A good program might fill one box. A complicated one could fill two.

I can still see Jimmy Salazar pushing his dolly beside me, his hair sticking up as if he had just stepped out of bed, which he often had.

"You think it will run tonight?" he would ask.

"It ran in my head," I would say. "That counts for something."

"It does not count for much with that machine."

He was right about that.

We would gather outside the computing center doors a few minutes before midnight, watching the big clock on the far wall. At twelve sharp, the temple allowed us inside.

The card readers stood like confessionals along one wall. You fed your deck into the slot and prayed the machine would pull them through cleanly. The sound was mechanical and final. Each card made a precise slap as it passed through. If a card jammed, the attendant would pull it out with a look that said you had offended something sacred.

"You're going to have to re-punch this one," he would say.

A mangled card meant your numbering was out of sequence. If you had been wise, you had written them in pencil on the back as well. If you had not, you would need last night's printout to reconstruct the program.

Once, making my way to the computing center, I hit a rock in the sidewalk near the engineering building. The dolly tipped. My box of punch cards slid off and scattered on the ground, fanning out like a flock of birds startled from a field. For a moment I just stood there, staring at them being wetted by early-morning dew.

Jimmy did not laugh. He squatted down and started picking them up.

"You numbered them, right?" he asked.

"I think so."

"That is not an answer."

We gathered the cards in silence. I took them back to my apartment and sat on the floor, restacking them by number. It took nearly two hours. I was done for that night.

Inside the computing center after midnight, there was an odd fellowship. We were competitors in the classroom, but here we were brothers in poverty. Nobody in my group had enough money for daytime runs. Nobody had enough sleep. We had equations to solve, data to reduce, and dissertations that depended on this machine accepting our offerings.

After feeding the cards into the reader, there was nothing to do but wait. The jobs would queue up. The 360 would process them in its own time. We could not see it work. We could not interrupt it. We could only retreat.

The only all-night café near campus was a narrow place with cracked vinyl booths and a waitress named Loretta who had stopped being surprised by physics students years before. We would push two tables together and order coffee we could not afford.

"You boys feeding the beast again?" she would ask.

"Yes ma'am," Jimmy would say. "Trying to earn our doctorates."

She would shake her head. "Seems like a lot of trouble for a piece of paper."

We would sit there until nearly three in the morning. Sometimes the talk was physics. Sometimes it was about girls or cars or the unfairness of professors who had forgotten what it meant to be poor. But always there was a clock in the back of our minds.

At some point someone would glance at his watch and say, "Let's go. We're probably getting output."

The walk back to the computing center felt longer than the walk from it. Anticipation has weight. We would step inside and scan the wall of output slots, each labeled alphabetically. You would find your last name. Benson. Sometimes the slot was empty, meaning your job had not finished. Sometimes there would be a thick stack. That meant success, or at least activity. Half the time, it meant failure.

I remember the first time I pulled a hefty stack from the B slot and flipped to the last page hoping for results. Instead, there was a block of capital letters.

ERROR 107.

I went to the shelf where the IBM 360 error manual lived, thick as a family Bible, and found my number.

107. INVALID DIMENSION IN ARRAY REFERENCE.

That was supposed to help?

Jimmy leaned over my shoulder. "What did you do?"

"I do not know."

"That is the problem."

We would sit there, tracing through pages of code printed line by line, hunting for the mistake. The machine did not forgive such things. It did not suggest. It condemned.

By the time we fixed the error, it might have been four in the morning. You could feed the corrected deck back into the reader, but you would not see results until the next night. You had to carry your uncertainty home with you.

There was an engineering student named Karen Whitfield who seemed to have a special touch with the 360. Her programs ran more often than they failed.

"How do you do it?" I asked her once.

"I imagine the machine is stupid," she said. "Because it is. It only does exactly what you tell it to do. Nothing more."

"That is not how it feels."

"That is because you are giving it too much credit."

She was right. The 360 was not cruel. It was literal. It executed instructions without mercy and without malice. The cruelty we felt was our own inadequacy, reflected at us in capital letters.

Some nights were worse than others. Three or four of us would all discover errors at once. The air in the computing center would thicken with frustration.

"This is ridiculous," Jimmy would say. "We are physicists. We can derive Maxwell's equations, but we cannot convince a machine to diagonalize a matrix."

"You are not physicists yet," Karen would reply. "That is the point."

I remember one run involving cosmic ray data. The stack filled nearly two boxes. I had checked it three times. I had walked the dolly as if carrying glass. When I found the output in the B slot, it was thicker than usual. That was a good sign.

I turned to the final page and saw numbers. Columns of them. No codes. No abrupt termination. Just results.

Jimmy slapped my back. "You did it."

"I think so."

"Do not think. Look."

We unfolded the pages across a table, scanning for anomalies. The numbers matched what I had calculated by hand for a small subset. They were not perfect, but they were plausible. That was enough.

We walked back to the café before dawn, not because we needed coffee, but because we needed to sit somewhere ordinary and let the tension drain away. Loretta poured without asking.

"Good news?" she said.

"For once," I answered.

She smiled as if she had seen this scene a hundred times, which she probably had. Students in the small hours, measuring their lives in stacks of output.

What strikes me most now is not the inconvenience, but the camaraderie. We were bound together by a machine that did not know we existed. We shared secrets about compiler options and card-numbering tricks. We lent each other rubber bands and pencils. When someone spilled a deck, nobody walked away.

There was humility in that ritual. You could not bluff the 360. You could not charm it. It did not matter how tired you were or how brilliant you thought yourself to be. It demanded precision, patience, and a willingness to begin again after failure.

Those nights shaped my understanding of work in ways that the lectures did not. When I think of those years now, I do not remember the exams. I remember the sound of a card dolly's wheels on concrete, the careful way I held a box of cards against my chest before feeding the reader, and the long walk to the café with students who were as uncertain and determined as I was.

We learned to stack our lives in order, to number the pages so that if they fell, we could gather them up and begin again.

And in that small temple after midnight, among the whir of card readers and the rustle of fanfold paper, we became, slowly and without ceremony, physicists.

CHAPTER 36

The Invitation

In the early spring of 1979, when the wind still carried a chill across the Texas A&M campus and the live oaks had not quite committed to green, Dr. Duller called me into his office. His office always smelled faintly of chalk dust and coffee that had gone cold. Papers were stacked in deliberate piles, not disorder, just layers of intention. He sat behind his desk, fingers tented, looking at me with that expression he wore when something important was about to be said.

"Robert," he began, "I have received an invitation."

I waited. Invitations in physics did not arrive lightly. They came through networks invisible to most of us, whispered across continents through letters and quiet recommendations.

"There is a summer school in Erice, Sicily. Cosmic ray physics. Very select. They have asked if I would send one of my graduate students."

I tried to keep my face neutral. Erice was not just a conference. It was a summit. In our small world of cosmic rays and particle showers, Erice had become something like myth. Names that appeared in journals, names that carried weight in faculty meetings and grant reviews, gathered there. Young scientists did not apply. They were invited.

He leaned back slightly. "It would be good for you."

I said nothing. I was afraid that if I spoke too soon, the spell would break.

"They do not teach classes in the usual way. You sit with the greats at long tables. You argue. You drink wine. You talk about ideas that might never be seen in print. No titles. No Dr. this or Dr. that. Everyone on equal footing. Even Bruno Rossi is expected to be there."

The name landed with force. Rossi. One of the pillars of the field. His work shaped the very problems we were trying to solve underground and on mountaintops. His papers were not just citations to us. They were foundations.

Dr. Duller studied me. "You would meet people who will shape the next twenty years of the field. They would know your name. They would hear your ideas. You understand what that means."

"I do," I said quietly.

"There is money in my budget to support your participation once you are there." He paused. "But not for travel."

I felt my stomach drop. He continued in the same measured tone.

"You would need to get yourself to Europe."

Europe. The word stretched between us like a distance I could not measure.

"I do not tell you this lightly," he said. "You need to find a way. This would not just be a pleasant trip. It could change your trajectory. These gatherings in Erice are rare. If you do not go now, you may not be invited again."

I left his office with the weight of it pressing on my chest. I did not have the money for a ticket to Italy. My stipend barely covered rent and groceries. I had a family. I had obligations. The idea felt cruel. To glimpse the summit and then be told I had no boots for the climb.

For several days I moved through campus in a fog. I checked data from the underground detector. I made notes in my lab book. I nodded when spoken to. But inside, I was calculating airfares, imagining numbers that did not exist in my bank account.

One afternoon I took the path I often walked between the physics building and the Cyclotron Building, cutting across toward Evans Library. There was construction fencing around part of the library. The expansion was underway. The old brick walls were being pushed outward, modernized, and made larger. Through a gap in the barrier, I could see bricklayers at work, their lines taut, their trowels flashing in rhythm.

I stopped. The sound of mortar being spread, the slap of brick against bed, the quick scrape of excess. It was a language I knew. My body remembered it before my mind did.

That used to be me.

I watched a mason butter a brick and set it into place with two confident taps. He checked the line without looking flustered. The work was physical, repetitive, honest. You could see what you had done at the end of the day.

The thought arrived whole. I did not need a grant. I needed wages.

I found a place where the barrier sagged slightly and slipped through. A man in a hard hat stood near a stack of brick, studying the alignment of a wall. His shirt sleeves were rolled up. His face had the weathered look of someone who had worked outdoors for decades.

"I need to see the masonry foreman," I said to a laborer.

The laborer pointed. "That's him."

I walked over. "Sir, are you the masonry foreman?"

He glanced at me, taking in the slacks, the absence of a hard hat. "I am. What do you need?"

"I'm a bricklayer. I was wondering if you might need another hand."

He looked at me longer this time. A hint of amusement crossed his face. "You're a what?"

"A bricklayer."

"You a student?"

"Yes."

He almost laughed, but not unkindly. "You don't look like any bricklayer I've hired lately."

"I worked several years before coming back to school," I said. "Commercial and residential. Block and brick. I've run lines. Laid corners. I can keep up."

He crossed his arms. "Where?"

I named the town. Described the contractor. I mentioned the kind of work we did, the long summer days, the piecework pace, the expectation that you did not slow the line.

He studied my hands. They were softer than they had once been, but the memory of calluses remained.

"You got tools?" he asked.

"Yes, sir."

He nodded once. "Be here at 7:30 in the morning. Bring your tools. I'll give you a try."

The next morning, I arrived before he did. The air was cool and damp. I carried my trowel, level, line blocks, jointer. When the crew gathered, he pointed to a stretch of wall.

"Start there. Let's see what you can do."

The first brick settled into mortar like it had been waiting. Butter, set, tap, scrape. Check level. Keep the line tight. The rhythm came back into my body. By midmorning sweat ran down my back and my shoulders burned in a way they had not in years.

The foreman watched without speaking. At lunch he nodded once. "You're all right."

That was praise enough.

For two months I lived in two worlds. In the mornings and afternoons, I laid brick. You would have laughed at how out of place I looked when returning to the physics building: hard hat on head, mud on my jeans, and my tool bag in hand.

At night I analyzed data and pondered cosmic ray showers. My hands grew rough again. My back ached. But each week I set aside money in an envelope marked Italy.

Sometimes the foreman would stand beside me and say things that stayed with me.

"Don't rush the corner," he said once. "The corner tells the truth about the whole wall."

In physics, the boundary conditions tell the truth about the model. I never told him that. He would not have cared. But I carried the sentence with me.

When I finally counted the money and bought the ticket to Zurich, it felt like lifting a weight from my chest. Zurich was cheaper than flying directly into Italy. From there I could find my way south. I handed the envelope of cash across the counter at the travel agency. When the woman slid the ticket back to me, I held it for a moment before putting it in my pocket.

The flight was long and disorienting. When I stepped into the Zurich airport, the air felt sharp and clean. Signs appeared in German. I bought a train ticket to Torino, the name spoken carefully at the counter.

The train slid through landscapes that seemed too orderly to be real. Villages with red roofs. Fields divided into precise rectangles. Mountains rising in the distance, their peaks still dusted with snow.

In Torino I met the small group Dr. Duller had arranged for me to join. Two graduate students and a professor from another university were loading a van with bags and crates of papers.

"You must be Robert," one of them said in accented English.

"Yes."

"Benvenuto," the other said, smiling.

We began the long drive south. The professor drove. The students debated a recent paper in Italian.

"Ma non è possibile," one insisted.

"Perché no?" the other replied.

They turned to me. "He says the shower profile cannot flatten like that."

"And you?" I asked.

"I say perhaps at higher energy it could. We will see."

We stopped at roadside cafes. Espresso in small cups. Bread and cheese were eaten standing at counters. The landscape shifted gradually from northern order to southern vibrancy.

As we approached Napoli, the traffic thickened. Scooters darted between cars. Laundry hung from balconies. At the port we drove onto the ferry. Entire train cars were being rolled aboard on tracks inside the hull. The scale humbled me.

We climbed to the upper deck as the sun began to set. The sky turned orange over the water. Napoli receded behind us.

The professor poured small cups of wine. "To Erice," he said.

"To Erice," we answered.

Overnight the ferry hummed. I slept lightly. At sunrise Sicily emerged from the haze, golden in the early light.

We disembarked in Palermo and began the climb toward Erice. The road narrowed and twisted. Olive trees lined up the hills. Stone walls marked boundaries older than my country.

"Quasi arrivati," one of the students said. Almost there.

Erice appeared above us, stone buildings pressed together along narrow streets, ancient and improbable. It felt less like a conference site and more like a place where ideas had always gathered.

As we drove the final stretch, I thought of the brick wall at Evans Library. Of the foreman's nod. Of the early mornings and sore muscles. Every brick I had laid felt somehow present in that van, part of the road that had brought me there.

We parked and stepped out into cool mountain air. Voices drifted through the stone corridors in Italian, English, German. Laughter. Argument. Anticipation.

The first morning I understood that this was not a conference in any ordinary sense. A man whose papers I had cited for three years stood at a chalkboard and drew a line representing a shower profile at ultra-high energy. He turned to the room and said, simply, that he did not know what happened above that line. That was the lecture. Everything after was argument.

I kept quiet for the first two days. I listened. I took notes. I watched how the senior physicists handled disagreement, which was openly, without the deference I was accustomed to in seminars. One evening at the long tables, the conversation turned to measurements at depth. Someone asked whether the muon flux at fifteen hundred feet had been reliably established. There was some skepticism in the room about underground detector data.

I had spent two years fifteen hundred feet underground with that data. I said so.

Heads turned. Not dismissively. Just attention.

"What were your systematic uncertainties?" someone asked. He was not unkind. He was precise.

I walked through it. The detector geometry. The event selection criteria. The stability of the electronics over long runs. The ways we had checked for bias. I had worried over those uncertainties for months. They were not strangers to me.

The room engaged. Questions came from three directions at once. Someone pointed out a potential correction factor I had not considered. Someone else defended my method. The Italian student from the van leaned forward and said that if my numbers held, they constrained a model he had been working on for a year.

I did not solve anything that evening. I did not emerge as a figure of consequence. But I had spoken from what I knew and been taken seriously. That was the transaction. That was what Erice offered.

The evenings had their own currency. We wandered the narrow streets in small groups, the stones beneath our feet worn smoothly by centuries of passage. Often, we climbed to Castello di Venere above the town, its foundations dating to the Norman period, built atop even older ruins. From that height the sea stretched wide below us. The sun would sink into it slowly. We stood there in silence more often than not.

Later, in a room below street level, a massive wine barrel stood against the wall, stocked with local Marsala. Glasses were filled and refilled. The light was low. Someone would begin describing an experiment. Someone else would counter. Then the topic would shift to family, to music, to politics. The distinctions of rank softened.

The van that had brought me from Torino carried me part of the way back down the mountain. In Napoli, my Italian companions dropped me at the train station. We shook hands. I promise to write, to send preprints, to meet again.

I stood for a moment watching the van disappear into traffic.

On the flight home from Rome, I stared at the wing cutting through clouds and tried to name what had changed.

I had not escaped my past. I was still the son of a violent bricklayer. That history would not evaporate over wine and theory. But I no longer believed that history defined the limits of my future. For years I had carried a quiet suspicion that I was attempting something beyond my rightful station. That suspicion had eroded in Erice.

The dissertation remained incomplete. The financial pressures would not dissolve because I had watched sunsets over the Mediterranean. There would be bills and deadlines and nights of doubt.

But there would also be the memory of stone streets and open arguments. The memory of standing at the edge of knowledge with others unafraid to admit what they did not know. The memory of having made my own way there, brick by brick, envelope by envelope.

I did not know then whether the letters Ph.D. would follow my name. What I knew was simpler and more durable. I had stepped into a room where the future of my field was being imagined, and I had belonged.

That knowledge would travel back across the Atlantic with me.

CHAPTER 37

What's in a Grade

If you have never been to graduate school, and especially not physics graduate school in the 1970s, you may not know that there were only two passing grades.

An A and a B.

A C was not "average." It was failure in slow motion. A D was administrative shorthand for packing your belongings. An F was a door slamming.

Exams were engineered so that no one could score perfectly. Professors said it openly. If someone made a 100, then the exam failed to measure the class. There must always be separation at the top. The mean hovered around 50 percent. Fifty translated to a C. A C translated to probation. Two C's and you were gone.

We all knew the numbers.

A 3.0 grade point average was survival. Below that, you did not return unless your advisor went to war for you. Some advisors did not.

There were stories. One student dismissed from the program had taken his own life. I never knew whether it was true. No one confirmed it. No one denied it either. In that environment, it did not matter whether it was true. It felt true.

Competition was not subtle.

One week a student missed class with the flu. When he returned, pale and coughing, he asked another student for the homework assignment he had missed. The list he received was wrong. Entirely wrong. He worked the problems faithfully, turned them in, and received zeros. The student who gave him the list later shrugged and said he must have misremembered.

I had never seen behavior like that as an undergraduate.

In that world, kindness could lower your rank.

I was not a star student. I knew it. I was not at the bottom either, but I lived in the narrow band where survival required vigilance. I did not have the easy brilliance of some of my classmates. I had work ethics. I was stubborn. I had a certain ability to endure. But I did not have effortless speed.

That mattered.

It was in that atmosphere that the young assistant professor of mathematics came wandering one afternoon.

I was in my office, door open, books and papers stacked in drifts across the desk. The hallway hummed with a low drone of fluorescent lights. A knock sounded against my doorframe.

"Are you Robert Benson?" he asked.

He was thin, dark-haired, early thirties, with the alert, slightly anxious expression of someone not yet tenured. He wore a sport coat that had seen better days and carried a folder under his arm.

"Yes, sir."

"I'm Dr. Adrian Markham. Mathematics." He stepped inside without quite committing to the room. "Do you have a minute?"

I gestured toward the chair.

He remained standing.

"I'm offering a course this semester, General Relativity. But under university rules, a class must have five enrolled students in order to make." He smiled tightly. "I currently have four."

I nodded. Everyone knew the rules.

"I've spoken with several physics students. They all say they're too busy." He paused. "I wondered if you might be interested."

General Relativity.

Einstein's field equations. Curved spacetime. The bending of light around massive objects. I had heard about these things in popular accounts, always looking through glass at something I was not yet equipped to touch.

I should have asked questions.

I should have asked how it would be taught.

I should have asked about prerequisites.

Instead, I said, "I'd like to learn general relativity."

He straightened slightly. "Then you'll enroll?"

I hesitated for perhaps three seconds.

"Yes. I will."

He extended his hand immediately. "Thank you. You won't regret it."

I would.

The first week was disorienting.

I expected tensors to apply to physics. Metrics written down for Schwarzschild spacetime. Perhaps derivations of gravitational redshift.

"Let M be a smooth manifold," Dr. Markham began.

He filled the board with definitions. Hausdorff spaces. Atlases. Charts. Transition functions.

By the third lecture he was proving the existence and uniqueness of the Levi-Civita connection.

The other four students, all mathematics graduate students, nodded, scribbled, and occasionally asked questions that began with, "In the proof of the previous lemma..."

I wrote down symbols and hoped comprehension would arrive later.

It did not.

By the second week we were computing Christoffel symbols abstractly. By the third week, curvature tensors.

The words "general relativity" appeared rarely.

It was Riemannian Geometry.

I began missing homework deadlines.

The first missed assignment was accidental. The second was discouragement. By midterm, I was lost.

The math students discussed problems in the hallway with ease.

"I think the sectional curvature vanishes if the metric is flat."

"Of course. That follows immediately."

Immediately.

I stopped turning in homework.

A foolish rationalization took hold of me. I had joined the class to help it make. Without me, it would not exist. I was doing him a favor. Surely, he understood that I was not one of his hand-picked mathematicians. Surely, he would not treat me as such.

I avoided his eyes in class.

At the end of the semester, a note appeared in my mailbox:

Please see me in my office.

I walked down the mathematics hallway with a tightness in my chest.

Dr. Markham gestured for me to sit.

He closed the door.

"I have calculated final grades," he said evenly. "I intend to assign you an F."

The word landed like a physical blow.

"An F?" My voice sounded thin.

"You have failed to submit the majority of homework assignments."

I could feel blood draining from my face. An F meant immediate dismissal from the physics program. There would be no probation. No argument.

"I-" I began, but nothing coherent followed.

He studied me.

"You understood the expectations," he said.

"I thought..." I stopped. What had I thought? That goodwill counted as credit?

He leaned back slightly.

"I will tell you what," he said. "There is a blackboard over there."

I turned. A full blackboard covered one wall of his small office.

"If you begin right now, I do not mean later this afternoon, I mean this minute, and work through every assignment you failed to submit, I will reconsider the grade."

"All of them?" I asked.

"All of them."

"That's weeks of work."

"Yes."

My options were simple. Leave and accept expulsion. Or stay.

I stood and walked to the board.

He handed me a stack of problem sheets.

"Begin with the first missing assignment."

I picked up the chalk.

The first problem involved computing the Riemann curvature tensor for a given metric. My hand trembled slightly as I wrote the metric components across the top of the board.

He returned to his desk and began grading papers.

The room was quiet except for chalk against slate.

When I stalled, he would say, without looking up, "You need to compute the Christoffel symbols first."

Or "Check your index placement."

Nothing more.

I worked until five that afternoon.

The next morning, I arrived at eight.

He nodded once.

I erased the board and continued.

Five days.

Eight in the morning until five in the afternoon.

I ate sandwiches at my desk in the physics building and returned immediately.

I computed covariant derivatives. I derived geodesic equations. I corrected signs. I erased entire boards when an index had been misplaced three lines above.

The math students passed by the open door and glanced in.

I did not look at them.

There was no room in my mind for embarrassment.

By the third day, something changed. The symbols stopped being enemies. The Christoffel symbols were no longer arbitrary clusters of indices; they were connections. The curvature tensor was no longer monstrous; it was a measure of how vectors failed to return to themselves after parallel transport.

I began to see the structure.

On the fifth day, late afternoon, I placed the chalk down.

"I believe that is the final problem," I said quietly.

Dr. Markham stood and walked to the board.

He read carefully. Occasionally he asked, "Why is this term zero?" I answered.

Finally, he nodded.

"You have done the work."

He returned to his desk and wrote something on a sheet of paper.

"I will assign you a C."

A C was failure in graduate physics.

He looked up.

"It will not count toward your degree. But it will not end your career."

Relief flooded me so quickly I had to grip the edge of the desk.

"Thank you," I said.

He met my eyes.

"You were not doing me a favor by enrolling in that course," he said calmly. "You were enrolling in a contract."

I walked out of his office exhausted beyond anything I had experienced academically.

It was the most focused I had ever been.

And I never again confused goodwill with work.

CHAPTER 38

No Glass Wall

The first computer I ever touched did not belong to me.

The PDP-11 at Southwest Texas State sat at the Computer Center and it was not sacred. It was just a machine, a machine with senses all over campus, senses in the form of terminals. They did not show photos or graphs or anything but letters and numbers. It spoke at least two languages: FORTRAN and COBOL. I think it spoke several other languages, too.

Somewhere in my San Marcos years, a program called Colossal Cave Adventure appeared on the terminals, and it spread through the student population the way a head cold spreads through a barracks. You typed commands, go north, get lamp, xyzzy, and the computer typed back, describing tunnels and treasures and a pirate who stole your things when you weren't paying attention. There were no graphics. No sound. Just words on a screen and a cave that went on longer than seemed possible. I lost hours on it that I did not have to lose. What got me was not the game itself but the way it pulled you forward, one more room, one more passage, one more attempt to figure out what that plough was supposed to do. It was the first time I understood, in my bones, how a machine could be designed to make you forget yourself. That probably should have worried me more than it did.

At Texas A&M the IBM 360 was larger, more powerful, and even more remote. It lived behind glass and policy. During the day, funded research groups purchased time on it. Graduate students without money were ghosts who appeared after midnight. We pushed our small two-wheeled dollies stacked with carefully ordered punch cards and prayed we would not stumble. A single dropped stack meant catastrophe. One card out of order and the program was nonsense.

The machine was tended by operators. You did not speak to it. You spoke to them.

You fed your deck into a reader and hoped no card tore. You hoped you had not mis punched one column. You hoped you had not reversed two lines of code.

The machine always felt like someone else's property.

I was learning to compute, but I was not learning to control.

That bothered me more than I admitted at the time.

I have always done best when my hands are on the tool. I like to understand the mechanism. I like to see cause and effect without mediation. With those large institutional machines, I was dependent on operators, on schedules, on funding, and on luck.

If something went wrong, I could not open the cabinet and look inside.

I could not touch the brain.

That lack of control sat in me like a splinter.

It seemed unfair that knowledge required permission.

Then one afternoon someone mentioned that a microcomputer club was forming at Texas A&M.

I did not know what a microcomputer was.

The word sounded small. Personal. Unofficial.

I went to the first meeting out of curiosity.

There were perhaps ten of us gathered in a classroom. Folding chairs. Chalk dust in the air. Most of us were graduate students from various departments. Five of them brought machines.

'Machines' is a generous word.

They were naked circuit boards laid carefully on tables. No cases. No polished panels. Just green boards, chips soldered in place, thin wires running like exposed veins. A microprocessor sat at the center. Around it, there is a scattering of supporting components. There was no keyboard. No monitor. Across one edge of the board ran a row of perhaps eighty tiny LEDs.

From the board hung a flat ribbon cable that led to a teletype reader.

Someone flipped a switch. A few LEDs flickered to life.

I leaned forward.

One of the students unrolled a long strip of paper tape. It had holes punched across its width in patterns. Five possible holes per row. Each pattern is a number. Each number part of a program.

He fed the tape into the reader and began pulling it through by hand. The reader made no sound. The holes passed over sensors. The code streamed into a memory chip mounted on the board.

When the tape ended, he pressed a small start button.

The LEDs flickered in patterns.

"That's a simple summation program," he said. "It adds a column of numbers and displays the total in binary."

The LEDs represented the binary output. Ones and zeros. On and off.

It was crude.

It was beautiful.

There was no operator. No midnight queue. No glass wall. The entire computer sat exposed on a table. If something did not work, you could trace it with your finger. The processor, the memory, the bus lines. Everything visible.

The user had complete and utter control.

I was mesmerized.

This was computing stripped to its bones.

Someone else demonstrated loading a slightly more complicated program. Someone spoke about writing code directly in machine language. Someone mentioned that kits could be purchased and assembled at home.

Kits.

The idea that you could build your own computer felt revolutionary.

I went home that night unable to think of anything else.

Technology was moving fast. Small companies were emerging, offering microcomputers for hobbyists and businesses. These were not toys. They were tools waiting for someone to imagine uses.

I always carried a small entrepreneurial streak. The Ph.D. was a long march through problems that would not announce their solutions on any schedule I controlled. This new world moved differently. You built something. You sold it. You saw the result the same week.

Within months I was consumed.

Two friends shared my excitement. One had modest financial backing. The other and I brought technical ability. We began discussing the possibility of opening a microcomputer business in the Bryan-College Station area.

We applied for a dealership with Ohio Scientific. They were building systems that used multiple processors. Our best-selling unit eventually carried three

CPUs: a 6502, a Z80, and an 8080. We named the business BKM Microcomputers.

We rented modest space and set up demonstration systems. We installed business software for bookkeeping and inventory. We wrote our own small applications for local companies.

The first real sale I remember was to a feed store owner outside Bryan. He drove up in a pickup with a cattle trailer behind it, a man who kept his books in a cardboard box and reconciled them once a year with a pencil and a prayer. I walked him through the system at our demonstration table, showing him how accounts could be sorted and totaled in seconds. He watched the screen. He did not say much. Then he pulled out a checkbook.

"How much for the whole setup?" he asked.

I told him. He wrote the check without flinching.

I had spent three years surviving on a graduate stipend that covered rent if nothing else went wrong. That check was more than I earned in two months of teaching physics labs. I folded it carefully and put it in my shirt pocket.

I understood then what the difference felt like. Not just the money. The directness of it. A problem existed. We solved it. A transaction occurred. There was no committee, no peer review, no waiting for a machine to process your request after midnight.

Customers came. More checks were written.

One of our programs encrypted documents. With a key, the file could be decoded. Without it, the contents were unreadable. Businesses liked the idea of privacy, even if the threat was mostly theoretical.

I was no longer a graduate student waiting outside the glass room. I was at the center of something new. I understood the hardware. I understood the code. If a board failed, I could diagnose it. If software malfunctioned, I could repair it.

Control had finally arrived.

Physics began to fade into the background.

I continued to sign up for research credit each semester. On paper, I remained a doctoral candidate. In practice, I was rarely in the lab. Data went unanalyzed. Draft chapters unwritten. Deadlines drifted.

I told myself I would catch up.

The business required attention. Inventory shipments. Customer support. Software revisions. Demonstrations. There were always urgent tasks.

Writing a dissertation felt abstract compared to soldering a circuit board or closing a sale.

Months passed.

I stopped attending departmental seminars. I avoided conversations about research progress. When colleagues asked how my thesis was going, I answered vaguely.

"Making progress."

I was not.

The more money the business generated, the less urgent the Ph.D. seemed.

I told myself that, if necessary, I could return to physics later. The degree was important, yes, but so was opportunity. Microcomputers were not waiting politely for me to finish curvature tensors.

The business consumed my energy completely.

Then one afternoon a note appeared in my departmental mailbox.

Please see me at your earliest convenience.

Dr. Duller.

Dr. Duller was not an emotional man. He was precise, methodical, and disciplined. He valued progress. He valued measurable advancement.

I had been avoiding him.

I stood outside his office door longer than necessary before knocking.

"Come in," he said.

His office was orderly. Books aligned. Papers stacked neatly. A large chalkboard behind his desk covered with equations from some recent derivation.

I sat.

He did not smile.

"How is your research progressing?" he asked.

I hesitated.

"I've been... occupied with some other responsibilities."

"I am aware," he said.

He folded his hands.

"I have received inquiries about your absence from the lab. Your data has not been updated in several months."

"I've been working on a business," I said carefully.

"A business," he repeated.

"Yes, sir. Microcomputers."

He regarded me without expression.

"And does this business occupy your full time?"

"It requires significant attention."

"And your dissertation?"

"I plan to return to it."

"When?"

I had no answer.

Silence filled the room.

"Robert," he said finally, "graduate school is not a hobby. The department invests resources in its doctoral candidates. There are expectations."

"I understand."

"Do you?"

His voice was not raised, but it carried weight.

"You are listed as a full-time student. You receive research credit. You use departmental facilities."

"Yes, sir."

"And yet you are operating a private commercial enterprise."

"It's... temporary."

He leaned back.

"Are you committed to completing your degree?"

"Yes."

"Then demonstrate it."

His gaze held mine.

"You have allowed yourself to drift."

The word stung because it was accurate.

"I believe in your ability," he continued, "but belief does not substitute for work."

He opened a folder on his desk. Inside were records of my enrollment, my progress, or lack thereof.

"You are approaching a point at which the department must decide whether you remain in good standing."

My stomach tightened.

"I don't want to lose my place," I said.

"Then you must choose."

He paused.

"I will be direct. This cannot continue as it has."

The room felt smaller.

"You cannot remain nominally enrolled while absent in practice. There will be consequences."

He closed the folder.

"I suggest you consider carefully what you want."

I left his office with a sensation I had not felt in years. Fear.

The microcomputers had given me control. They had given me money and momentum and a sense of mastery. But now that same choice threatened to undo everything I had worked toward in physics.

I stood in the hallway outside his office and understood, perhaps for the first time, that control in one arena can cost you control in another.

CHAPTER 39

The Ultimatum

Physics departments do not generally award master's degrees as a destination.

They accept students directly from a bachelor's degree into the Ph.D. track and expect them to continue straight through. A master's in physics, though respectable on paper, often carries an unspoken implication. It can suggest that the student did not quite clear the higher bar. A consolation. A soft-landing prize.

Most physics doctorates do not hold separate master's degrees. They go in once and come out once.

The average time from bachelor's degree to Ph.D. in physics is about six to seven years. Six is efficient. Seven is acceptable. Eight years invite questions.

I had moved far beyond acceptable.

By late 1984 I had been in the program nearly a decade. My diversion into microcomputers had not merely slowed me. It had nearly derailed me.

Then the letter arrived.

It was formal, typed, unadorned. If I remained enrolled in the doctoral program for ten years, I would be required to retake all first- and second-year graduate physics courses before being allowed to proceed.

Retake everything.

Quantum mechanics. Electrodynamics. Statistical mechanics. Classical mechanics. The entire gauntlet.

I sat at my desk reading the letter twice, then a third time. I was approximately ten months from that ten-year threshold.

Ten months.

I was not writing my dissertation.

The words on the page felt like an explosion. Not loud, not dramatic, just concussive. A pressure wave that rearranged the air in the room.

Retake everything.

I tried to imagine myself sitting again in those early graduate classrooms. Facing those exams again. Competing again with younger students who had not spent years running a business on the side.

It was not merely the inconvenience. It was humiliation.

That letter removed all ambiguity. I could not continue as I had been. I could not be half physicist and half entrepreneur.

I drove to the BKM Microcomputers office that afternoon with a tightness in my chest.

The store smelled faintly of warm circuitry and paper manuals. Boxes of Ohio Scientific hardware lined the back wall. The demonstration unit sat on the counter, its LEDs glowing softly. A customer was browsing software catalogs.

My partners were in the back room.

"We need to talk," I said.

They looked at me and knew from my face that this was not a routine conversation.

"I got a letter," I began. "If I hit ten years in the program, I have to retake everything."

One of them whistled softly.

"How close are you?"

"Ten months."

Silence.

"You can do both," one said. "You've been doing both."

"No," I said. "I haven't. I've been pretending."

The words surprised me as they left my mouth, but they were true.

"I have to choose."

We sat around a small metal desk, the hum of a cooling fan filling the pauses.

"You built this," my partner said. "You don't have to walk away."

"I do," I answered.

There was money to be made. There was momentum. The microcomputer market was still young. We were first in the area. That mattered.

But the Ph.D. mattered too.

It was not just a degree. It was a line in my life that had begun years earlier when I first walked into a physics classroom and believed I could belong there.

I negotiated a sale of my shares to my partners. The terms were fair. There were handshakes. There was an awkward attempt at optimism.

"You'll be back," one of them said.

I nodded, though I did not know whether that was true.

The next morning, I returned fully to physics.

There was no gradual easing in. The data had long since been collected. Cosmic ray scintillations in muons near sea level. The apparatus had worked. The measurements existed. The analysis had begun, then neglected. I returned to it with a focus I had not previously applied to anything academic.

I rose early. I worked until exhaustion. I rewrote sections repeatedly. I rechecked calculations. I verified plots. I scrutinized error bars.

The urgency the business had trained into me, now transferred to my dissertation.

Weeks blurred into months.

Draft chapters circulated among committee members. They returned with red ink and marginal notes.

"Clarify this assumption."

"Expand this derivation."

"Justify this approximation."

It was humbling.

Each correction felt less like criticism and more like chiseling stone.

I learned to accept the edits without defensiveness. The dissertation was not a personal diary. It was a document that would carry their signatures.

When a doctoral committee approves a dissertation, they are not merely certifying that the student completed a task. They are attaching their own reputation to my work.

If I passed the final oral examination, I would walk into the physics world bearing their names beneath mine. If I performed poorly in my career, some shadow of that failure would reflect on them.

I understood this now in a way I had not earlier in my graduate life.

I was not only representing myself.

I was their product.

The writing became almost maniacal. A narrowing of focus that excluded nearly everything else. Friends noticed my absence. Conversations shortened. Evenings disappeared into revisions.

The completed dissertation, bound and formal, felt heavier than its pages should allow.

An Investigation of Cosmic Ray Scintillations in Muons Near Sea Level.

My name is beneath it.

Robert Henry Benson.

The committee members reviewed the final draft. There were more edits, substantial ones in places, but eventually the signatures came. Their names on the approval page transformed the stack of pages into something official.

But approval of the written document was only part of the trial.

The real test remained.

The public presentation would precede the closed oral examination. I would stand before the faculty and students and explain my work. Then, after the audience dispersed, the door would close.

In that room, the committee would sit.

No audience. No applause. No escape.

They could ask anything. Not just about my dissertation. About quantum mechanics. Electrodynamics. Statistical mechanics. Classical theory. Anything they deemed relevant to judge whether I deserved the title.

On the night before the examination, I sat alone with the bound dissertation on my desk.

I ran my hand across the embossed title.

Ten years.

Ten years that had nearly unraveled.

Ten years that now converged toward one closed door and a table of five men.

There would be no business to retreat into. No alternate identity. Only questions and my answers.

CHAPTER 40

Is All Lost?

At one o'clock on a Monday afternoon, I stood at the front of a classroom and gave the best thirty minutes of my academic life.

The slides clicked forward cleanly. The graphs were sharp. Underground muon counts, power density spectrum, slope of minus 1.36, the salt mine at fifteen hundred feet. I had rehearsed it until the words came without effort, and that ease freed something in me. I heard my own voice as if from a short distance away, steady and unhurried, the voice of a man who belonged at the front of a room.

When I finished, a student raised his hand.

"What was it like down there? In the salt mine?"

"Dark," I said. "And quiet in a way that has weight to it. You can hear your own breathing as if it belongs to someone else."

A little laughter. A few more questions, answered cleanly. The faculty nodded. No one tried to gut me in public.

We filed downstairs.

The conference room in the basement of the physics building was smaller than I remembered. A long table. A chalkboard along one wall. The smell of chalk dust and coffee that had gone cold some hours ago. Five men arranged themselves on one side of the table. I sat across from them alone.

Dr. Duller began. His questions came from the list he had given me, and I had lived inside that list for weeks. Derivations I had rehearsed until they arrived without effort. Assumptions I could state in my sleep. He nodded as I answered, and something in his expression told me I was performing as he had hoped.

The others took their turns. Tribble on statistical uncertainty. Green on calibration. Fahlquist on boundary conditions. I handled each of them. The room felt warmer as it went on, not from heat but from momentum. I could feel the end of it approaching like a clearing visible through trees.

Dr. Duller glanced at his watch.

"Well," he said, "I believe we are nearing the end. I'll ask each of you if you have any final questions."

Tribble shook his head. Green said he was satisfied. Fahlquist nodded once.

Duller turned slightly.

"Dr. Katawar. Any final questions?"

Katawar had been quiet most of the time, leaning back in his chair with his fingers folded across his stomach, looking half asleep. Now his eyes opened fully.

"Ah. Yes. I do have one."

He leaned forward.

"Mr. Benson, please stand at the chalkboard one more time."

The clearing vanished. Trees closed back in.

He folded his hands on the table and looked at me with the calm of a man who has all the time in the world.

"Please derive an equation for the scale height of the atmosphere of the planet Venus, from the planet's surface to the top of the cloud cover."

For a moment the room made no sound.

"I'm sorry?" I asked.

He repeated it. The same words, the same calm.

I walked to the board. The chalk felt unfamiliar in my hand, as if I had never held one before. I stood there looking at the blank slate and felt the ground shift beneath me the way it had shifted on a dark road outside Brownwood when ice appeared without warning, the way it had shifted in a salt mine when the lights went out, the way it had shifted in a hundred moments across a life built from moments exactly like this one.

I could not find the first stone to stand on.

Behind me, five men waited.

Ten years collapsed into that silence. The scaffold in the August heat. The block wall at Southwest Texas State. Deco on his porch steps with his red hair and his joint and his lazy grin. The old house on North Street. Midnight card runs to the IBM 360. The fluorescent tubes converted to Geiger counters. The flask exploding in my hands. The brick wall at Evans Library, and the foreman's nod. The ferry crossing to Sicily, and wine poured in a stone cellar, and Bruno Rossi arguing at a long table as if the answer mattered more than the man who found it.

All of it present in the chalk I was not yet writing with.

"I remember this from your class," I said. My voice came out steady, which surprised me. "But I'm sorry. I cannot recall how to get started."

Katawar stood and walked to the board. He took the chalk from my hand without ceremony, and wrote:

$$\frac{\partial P}{\partial z} = -\rho g$$

He stepped back.

"This is the hydrostatic equilibrium equation. You recognize it."

It was not a question, but I answered anyway. "Yes."

"You should be able to go from here."

He returned to his seat.

I stared at the equation. The symbols held still on the board but moved in my mind, rearranging themselves without settling. My heart was loud. The room was louder with its silence. I thought of the boy who had poured ink on a girl's head because he had no other language for his rage. I thought of the boy who had held his breath in a highchair until the room tilted. I thought of every moment I had stood in front of something larger than myself and discovered I was still standing when it was over.

I wrote something. It was not much, but it was a beginning.

Katawar spoke again, his voice patient as geology.

"Think about the ideal gas law. Mass. The universal gas constant. That should get you to the next step."

I wrote the ideal gas law beneath the first equation. My hand had stopped trembling.

"Good. Now eliminate density."

I eliminated density.

"Take the first derivative with respect to z."

I took it. Or tried to. The algebra that would have come easily a week before arrived now in fragments, each piece requiring effort I could feel in my shoulders. Katawar prompted at each stall. Not generously, not warmly, but steadily, the way a man leads another man through fog by keeping one hand on his shoulder and saying: this way. This way. This way.

I do not know how long it took. The board was filled with half-formed expressions and partial integrations. At some point the general shape of the derivation appeared imperfect, incomplete in places, but recognizable. Chalk dust had settled on my sleeve.

At last Dr. Duller cleared his throat.

"I think that will be sufficient."

He looked at me.

"Mr. Benson, please step outside while we deliberate."

The hallway was cool and very quiet.

A few graduate students stood nearby pretending not to wait. One of them looked at me. I shook my head and dropped my eyes on the floor. I put my back against the cinderblock wall and let it hold me up, because my legs were not entirely sure they wanted to. Ten years. Every early morning and every late night, every card deck dropped on a wet sidewalk, every draft chapter bled red by a committee member's pen. It could all come down to one man's vote. I had not given a clean derivation. Katawar had led me through it like a man leads another man through fog — one hand on the shoulder, one step at a time. Whether that was enough, I did not know. The door stayed closed.

I thought about Otto Galvan. About the way he had moved along a scaffold at the end of a day, his knee destroyed by decades of lifting, his face wearing the patience of a man who had long ago made his peace with difficulty. About the tamales his wife made, and the way he had explained them to me on a lunch break in the heat of an August that felt like another life.

'You got to taste some bad ones to know the good ones.'

The conference room door stayed closed.

I had not given a perfect derivation. I had not stood at that board like a man who knew. I needed help at every step. I had been led through the fog by a quiet voice and a single equation written in another man's hand.

The door opened.

Dr. Duller stepped into the hallway. He walked toward me with his hand extended and his face holding something I had never seen in it before. Not a smile exactly. Something quieter than that. Something that looked like satisfaction.

"Congratulations, Dr. Benson."

The others filed out behind him. Handshakes. A few words, kind and brief. Katawar came last. He clapped me once on the shoulder, and his face had broken open into a grin wide enough that I barely recognized him.

"Hey, Benson. You were getting off too easy. I wanted you to remember this day."

He held out his hand.

"Congratulations, Doctor."

I walked out of the physics building into the late afternoon. The Texas sun was at the angle where it made everything look briefly valuable, casting long shadows across the quad, lighting the edges of things.

Students crossed the grass unaware. Couples walked. A dog angled across the lawn at a trot, following something invisible. The ordinary world had continued exactly as it always did, indifferent to the fact that something inside the building behind me had shifted.

I stopped on the sidewalk and stood still for a moment.

I thought about scale height. About the way atmospheric pressure falls with altitude, exponentially, thinning with each upward foot until the weight of the world becomes almost nothing. About Venus, its atmosphere is so dense at the surface that it crushes whatever lands there, its clouds so high that from above they look peaceful and white. About the distance between those two places. About what it costs to rise from one to the other.

I had spent my whole life learning that distance.

The low points had not been obstacles. They had been the measure. Without the scaffold and the heat and Otto's ruined knee, the word *Doctor* would have been just a word. Without the ink poured on a girl's head in third grade, without the cold water in the sink, without the belt and the highchair and the turkey massacre and the flask exploding in my hands and the years I spent underground counting particles in the dark, without all of it, there would have been nothing to rise from.

You cannot know the good ones without the bad ones.

I started walking toward the parking lot. The sun dropped another degree. The shadows stretched longer across the grass. Somewhere on the other side of campus a bird called once and went quiet.

I did not look back at the building.

I had what I came for.

Epilogue

It has been a long journey since Otto, and I climbed down that scaffold and shook hands goodbye.

If you have read this far, I hope you have sensed the theme that runs through my story. My life has rarely been lived in the middle. Instead, it has swung between deep lows and exhilarating highs. Over time I came to believe that a life lived at those extremes, though often painful, can be richer than one lived safely in the center. It is the distance between the highs and the lows that gives life its color.

Otto hinted at that truth long before I fully understood it. When he bragged about his wife's tamales he liked to say, "You can't appreciate the good ones if you haven't tasted some really bad ones."

At the time I laughed, thinking it was just a joke about food. But over the years I realized that simple comment contained more wisdom than I understood then.

Still, a life lived only at those emotional extremes cannot continue forever. Even the wildest storms eventually need calm water.

Around the time I graduated in 1985, I met Karen Lynne Partridge. We fell in love and got married. From the beginning we shared many passions, especially our love of the outdoors and the natural world. Yet in temperament we were very different, and that difference turned out to be our strength. Karen was, and still is, the calm that steadied my storm.

Left to myself I might have continued riding those radical emotional swings until they eventually broke me. Karen cast a quiet net over our lives and gently reined in my extremes. Not completely, and I would not have wanted that, but enough to keep us moving forward.

Soon after graduating, I was fortunate to be hired to fill a suddenly open faculty position at Texas A&M University in the College of Engineering. It is rare for a major university to hire one of its own graduates as faculty, and I knew how lucky I was.

But becoming an assistant professor is not the comfortable landing many people imagine. A new professor is given six years to prove his worth. During that time, he must teach effectively, bring in significant research grants, and publish his work in respected scientific journals. At the same time, he must earn the respect and confidence of his colleagues.

At the end of those six years comes a moment of judgment. If he succeeds and is awarded tenure and promoted to Associate Professor, he continues his journey. If he fails, he packs his office and starts over somewhere else or at another job outside academia.

I was able to meet those demands and earned tenure, beginning the next long climb toward the rank of Full Professor. That stage requires something different. You must help guide younger faculty and become widely respected in your field of research.

My path toward that recognition unfolded in a way I could never have predicted when I first entered physics.

Throughout my life I have been drawn to wild places and wild creatures. At the same time my training in the physical sciences has given me tools for analyzing complex signals. At some point those two worlds simply came together.

Birds sing songs.

Those songs are sound waves. Sound waves can be recorded, measured, and analyzed using the same instruments and methods I had learned in physics.

Out of that realization a new direction opened for me. My field became bioacoustics, the scientific study of biological sound.

At TAMU, I founded the Center for Bioacoustics, and through that organization I spent my career capturing and studying the voices of the natural world. My students and I recorded whales in the Gulf of Mexico, pink dolphins in the Amazon River, birds across Texas, Mexico, South America, and Europe.

Along the way, I had the privilege of working with many brilliant graduate students who came to study in the Texas A&M System. Watching those young scientists grow into accomplished researchers became one of the most rewarding parts of my career.

Karen shared this adventure with me. For more than forty years we have lived what I can only describe as our dream lives. The boy who once wandered fields and woods listening for birds somehow grew into a man whose job was to travel the world listening to and studying the sounds of nature.

Looking back now, I can see that none of it would have happened without the strange combination of forces that shaped my life. The turmoil of my early years pushed me forward with restless determination. Karen brought the stability that allowed that drive to last a lifetime. And above everything, was the unpredictable role of luck.

If I had the chance to live this life again, I would change very little. Only this: I would try to be kinder to the people I hurt along the way, whether through impatience, selfishness, or simple blindness to the pain my actions may have caused. Aside from that, I would gladly live it again much the same way.

Sometimes I think back to that day on the scaffold with Otto. We climbed down at the end of the workday, covered in dust and mortar, and shook hands before going our separate ways. Neither of us had any idea what paths lay ahead.

At the time his remark about tamales seemed like nothing more than a bit of humor. But over the years I came to realize he had unknowingly described the shape of a life.

The bitter years make the sweet ones shine brighter. The failures make the victories feel almost unreal. The lows give meaning to the highs.

And when I look back across the long arc of my life now, I realize something Otto himself may never have known.   You really do have to taste some bad ones before you know how good the good ones can be.

Photographs

My War-time father, James Albert Benson

James Benson fighter (US Army)

Florine Palmero 1950 (my mother)

Robert Henry Benson (My favorite shirt)

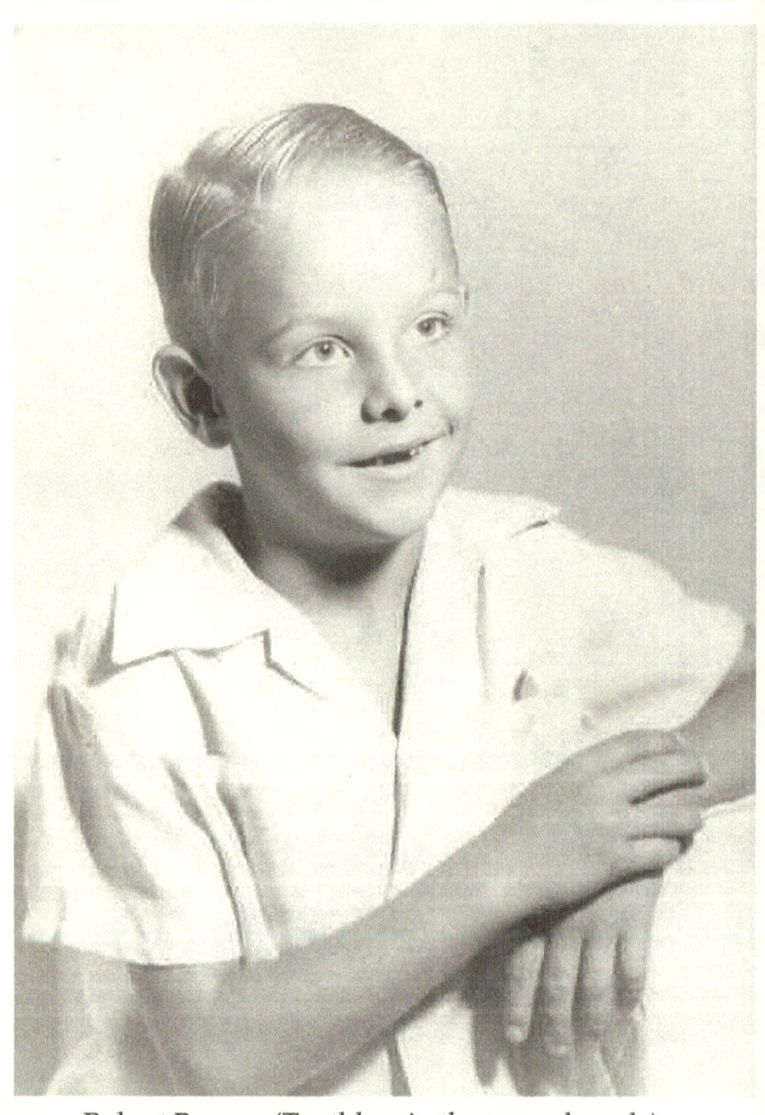

Robert Benson (Toothless in the second grade)

Robert Benson (Where I was sent because I couldn't learn)

Robert Benson (River Oaks Living)

Robert Benson (The Year I dropped out of school)

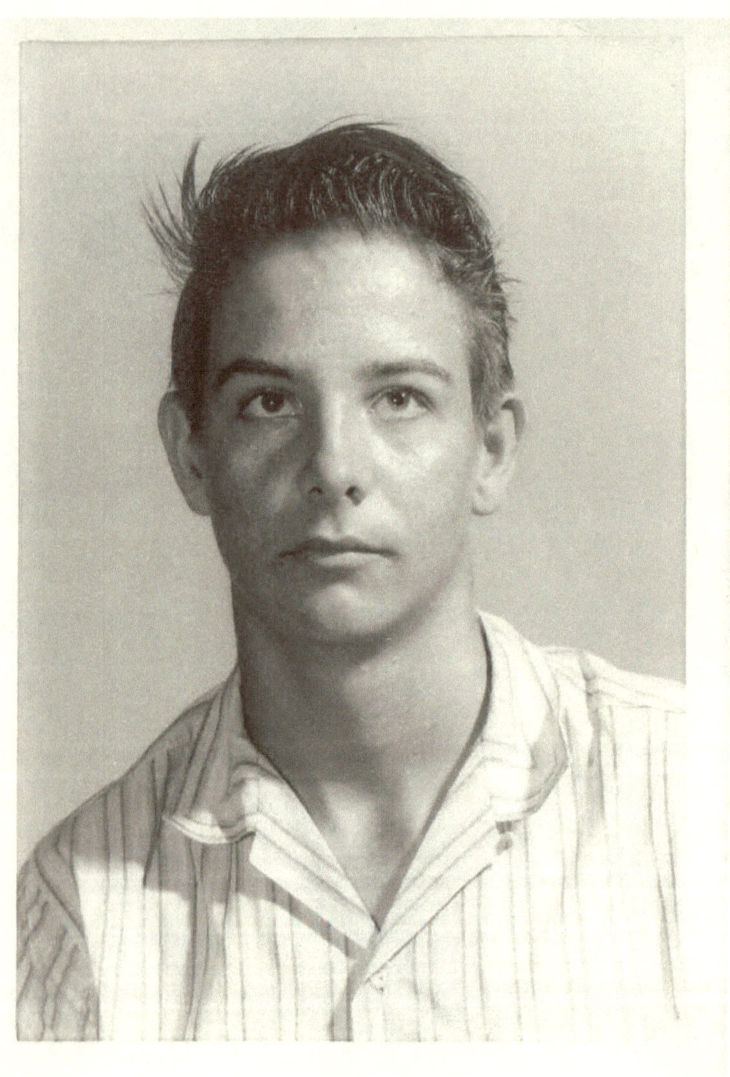

Robert Benson (Trouble ahead)

Rachel Smith and Robert Benson with our daughter
Tracy

Robert Benson (L) and John Benson (R), the telescope
that caused the deadly green gas

Deep in the Heart of Texas – 1,500 feet deep to be exact – Texas A&M physics professors measure muons (cosmic rays) as part of a research project to study variations that may be attributable to large-scale magnetic fluctuations as well as other changes in our galaxy. The cavern is part of the United Salt Corporation's mine in Hockley, Texas.

4   THE TEXAS AGGIE   FEBRUARY 1978

Robert Benson (far left) 1500-feet underground in a salt mine near Houston, Texas

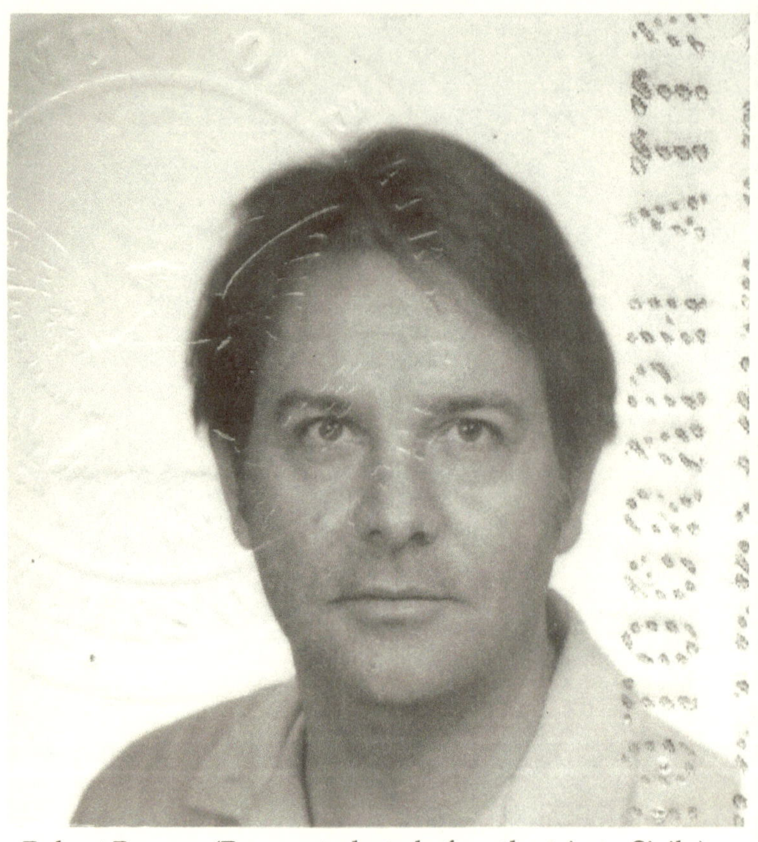

Robert Benson (Passport photo before the trip to Sicily)

Dr. Robert Benson (hand wave front row) 1985
commencement TAMU College Station

**IF YOU ENJOYED MY MEMOIR, PLEASE CONSIDER
POSTING A REVIEW ON AMAZON.COM
THANK YOU**